WITHDRAWN

The *The* HUBBLE

SPACE TELESCOPE

The
HUBBLE
SPACE TELESCOPE
Imaging the Universe

David DeVorkin

Robert W. Smith

NATIONAL GEOGRAPHIC

WASHINGTON, D.C.

28.5° N

28.5° S

Sample section
of Hubble orbital
path

The Hubble Space Telescope (HST)
circles the Earth's surface every
97 minutes at an average altitude
of 375 miles (600 km) and an
average speed of 17,500 miles
(28,000 km) an hour.

Contents

BOLDFACE indicates date, where available, that image was released to the public.

PG 1: **12.25.99** HST being returned to orbit from *Discovery*.

PP 2–3: **4.24.90** The 24,500-pound Hubble, here in orbit, is the size of a bus.

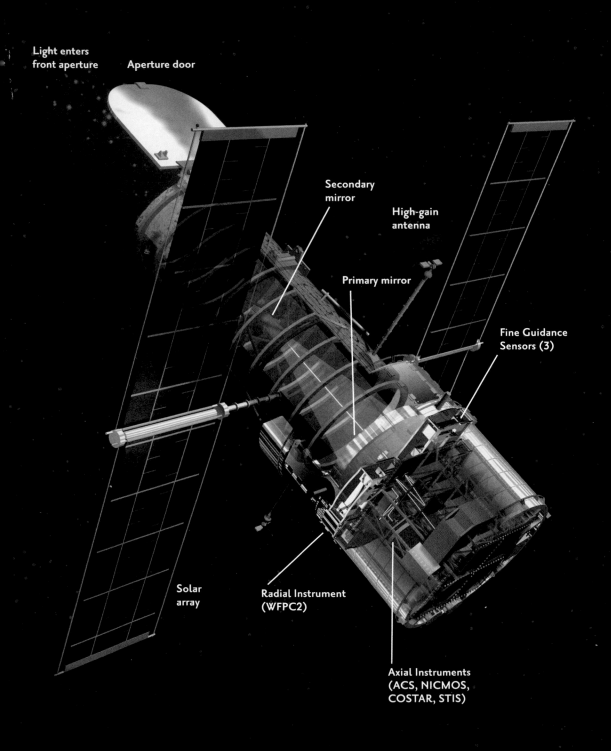

Light enters
front aperture

Aperture door

Secondary
mirror

High-gain
antenna

Primary mirror

Fine Guidance
Sensors (3)

Solar
array

Radial Instrument
(WFPC2)

Axial Instruments
(ACS, NICMOS,
COSTAR, STIS)

A T THE MENTION OF THE HUBBLE SPACE TELESCOPE, CHANCES ARE THAT SOME RATHER REMARKABLE IMAGES COME TO MIND. THE NAMES OF SOME OF THESE ARE GENERALLY FAMILIAR: "THE EAGLE NEBULA" OR "HORSEHEAD" OR "THE MICE." THE HUBBLE SPACE TELESCOPE MAY WELL BE THE MOST WIDELY KNOWN SCIENTIFIC INSTRUMENT IN HISTORY. THIS IS SO LARGELY because of the impact of these images, some of which have reached far beyond the scientific literature to become a familiar part of our modern cultural landscape.

Hubble is a highly sophisticated multifunctional observatory. It is an incredibly complex robot that is continually instructed by its human controllers to view the universe in a multitude of ways, and then to send this information back to Earth for analysis by a vast array of humans and machines. Within these pages is an overview of the extraordinarily rich history of imaging with Hubble, of techniques chosen to do the imaging, arguments made for imaging as Hubble was being planned and built, the crisis of the flawed main mirror after launch, how it was fixed, and how improved light detectors have sharpened and deepened its view. The Hubble in orbit now is a much more powerful scientific tool than it was at its launch into space in 1990. We will follow the life of an image from conception to publication, finding out how astronomers gain access to the Hubble, how Hubble acquires objects, collects light from these objects, senses this light, stores this light, and transmits the information it has secured back to Earth.

To place the Hubble Space Telescope in a historical context, the motives and drives of observational astronomy since the invention of the telescope 400 years ago will be narrated. Hubble is the result of a lavish investment of effort, time, and money to better understand our universe, but is also an observatory that is based on a long history of invention, ingenuity, and innovation in the construction of path-breaking telescopes.

Beyond the scientific goals, the book will also show how a small band of astronomers persuaded the Space Telescope Science Institute to form the Hubble Heritage Project to create compelling images from scientific data already accumulated and to release these striking images to the general public. All sorts of aesthetic judgments are made in shaping the imagery; color, contrast, and framing generate a sense of the infinite and sublime. The wonder of the universe has never been as compelling as it is today thanks to the Hubble Space Telescope.

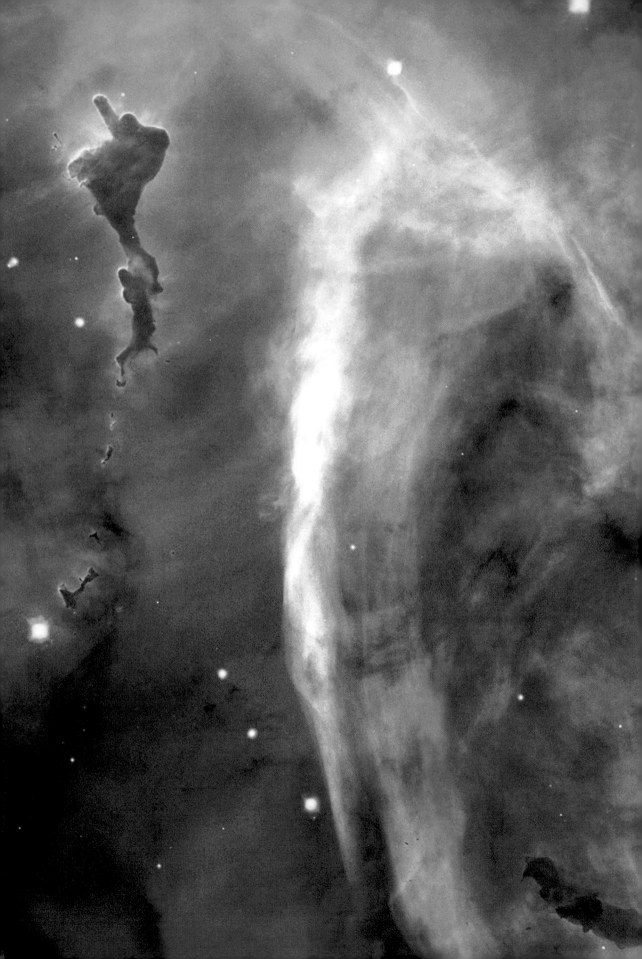

The Realm of the Nebulae

This chapter's gallery showcases galaxies and nebulae, subjects of intense debate even before the 20th century.

2.03.00 Portion of the Keyhole Nebula in the constellation Carina.

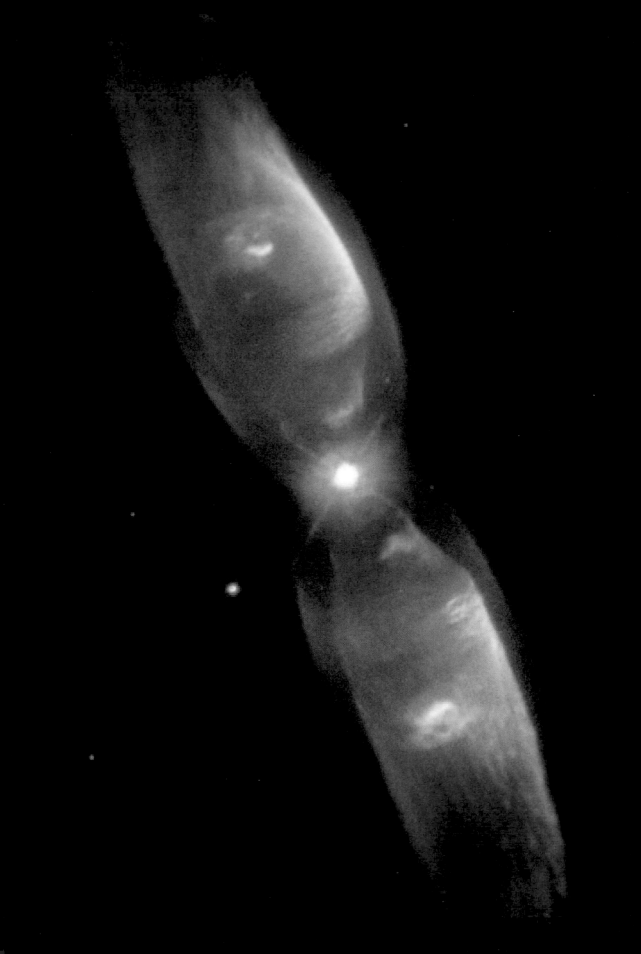

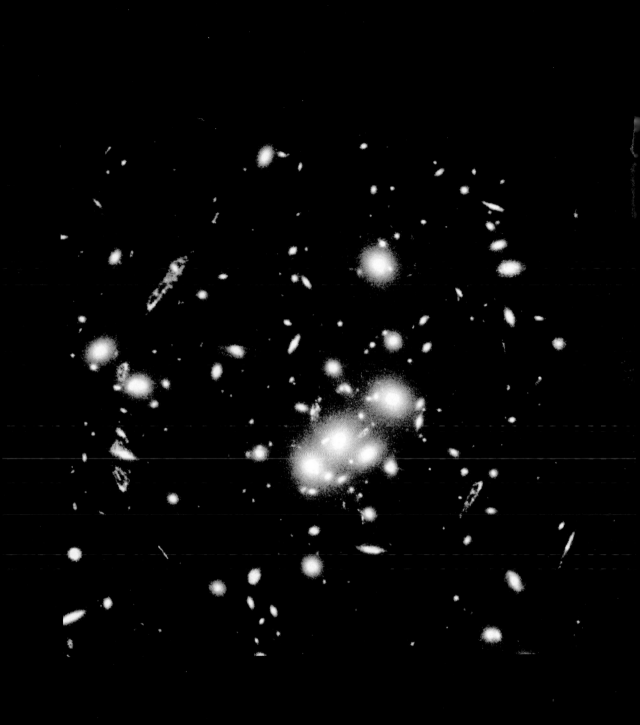

LEFT: 12.17.97 A bipolar planetary nebula in Ophiucus evokes a butterfly.

ABOVE: 4.24.96 Light distorted by dark matter in a galaxy cluster.

The southern Milky Way, as
depicted over London in the 1860s.

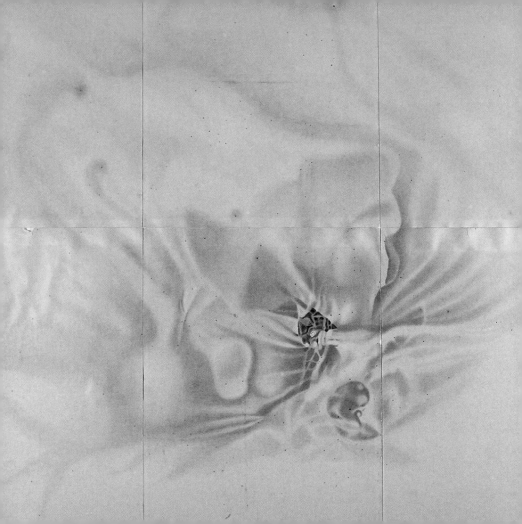

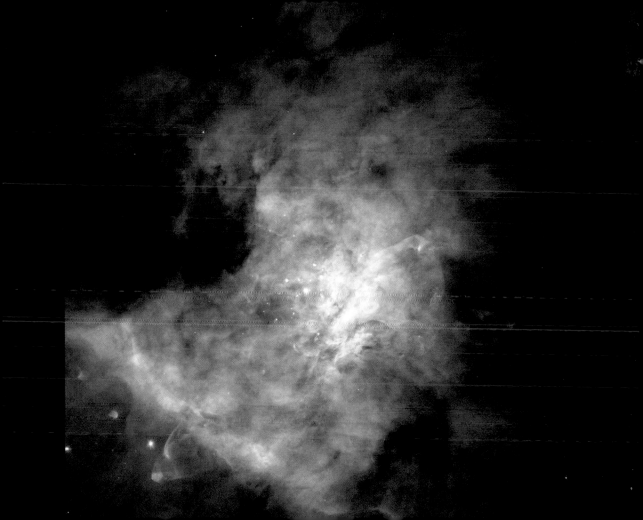

Ground-based image of a star in the
Fox Fur Nebula.

7.19.01 Galaxies in the process of collision in Stephan's Quintet.

The Realm of
the Nebulae

JUST AFTER SUNSET ON A SUNDAY IN EARLY JANUARY 1610, A MIDDLE-AGED PADUAN ASTRONOMER TURNED A THIN TUBE-SHAPED DEVICE TOWARD JUPITER, RISING IN THE SOUTHEASTERN SKY IN THE CONSTELLATION OF TAURUS, THE BULL. HE HAD BEEN BUILDING AND USING THESE DEVICES FOR ALMOST SIX MONTHS, BUT HAD TURNED THEM TO THE SKY ONLY RECENTLY.

The astronomer was excited by the sights they had disclosed. Through them he had seen that the moon was not a perfect orb, as Aristotle taught, but was riddled with pocks, mountains, and seas: It was Earthlike. He had been examining the moon since late November, and now he turned to Jupiter to see what his perspiculum, or spyglass, would reveal. Jupiter was a tiny disk, much smaller than the moon, but definitely not starlike. He could, however, make out three tiny stars in a perfect line drawn through Jupiter, one to the west and two to the east. The next night, "guided by what fate I know not," he again trained his glass toward Jupiter. The three stars were now all on one side of the planet! He continued to observe Jupiter and these stars every clear night, musing how they seemed to mimic the motions of tiny planets around a central sun. After a week he began to notice actual movement from hour to hour. Each night he carefully drew what he saw and recorded these images together in a manuscript he was preparing for publication. This record of his little stars and their motion around Jupiter became "possibly the most exciting single manuscript page in the history of science."[i]

Galileo Galilei, Chair of Mathematics at the University in Padua, did not invent this astonishing device, but he did improve its ability to magnify. A convex lens at the front of his spyglass, or optick tube, or—as it was soon coined—his telescope collected and concentrated light from a distant locale in a thin converging cone to intersect the smaller concave lens at the back of the tube where the light then converged into his eye. Galileo's first telescope magnified objects three times, making them appear three times closer and so nine times larger. Soon he was making telescopes that could render objects 30 times closer. [ii]

Galileo and others realized that telescopes possessed additional capabilities. Dim objects appeared brighter (we call this "light-gathering power") and could clarify indistinct or blurry scenes (what is called "resolving power"). The advantages of such a glass for the identification of friend or foe at sea or on land was obvious. But Galileo was the first to use the telescope for sustained and systematic study of the heavens. In doing so, he applied considerable talent as an observer, well versed in "the art of picturing and drawing."[iii] He saw the universe through new eyes and sought to share these stunning and novel views with others.

Galileo knew the persuasive power of visual evidence and convincing argument. He established the power of the telescope to penetrate into the depths of space, showing that celestial objects were not what they seemed to the naked eye. The first public announcement of his remarkable findings in the heavens came in a small book titled *The Starry Messenger,* published hastily in 1610. In it he included drawings of the moon and the motions of the Jovian moons. Galileo also explained how, viewed

Galileo Galilei,
portrait by Sustermans

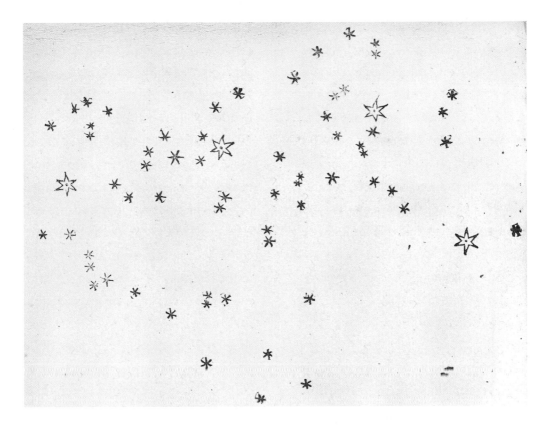

Portion of Orion's Belt and Sword from Galileo's *Sidereus Nuncius*.

through a telescope, the Milky Way revealed its true nature as a vast realm of faint stars. He drew the star fields of the Belt and Scabbard of Orion and the Pleiades and Praesepe, and described how, with the aid of his telescopes, he could detect far more stars in these fields than were visible to his eyes alone.

Galileo's success at resolving the Milky Way into stars, and showing that fuzzy patches in Orion and the Pleiades were in fact fields of stars too faint and densely packed to be seen individually with the eye, mark the beginning of a 300-year quest to

picture the universe as an assemblage of stars. Even as the powers of telescopes were constantly improved, however, this quest was often challenged by the astronomers' detection of blurry nebulous objects. Many of these misty objects were shown by the 1920s to be vast systems of stars lying far beyond the boundaries of the Milky Way, what we today call galaxies. The attempts to determine the nature of the nebulae stimulated the practice of telescopic astronomy, the drive to represent the physical nature of the objects in the universe through naturalistic rendering. This enterprise has engaged astronomers using the most powerful telescopes that patronage and technology could provide. Along with this growth has come an evolution in the means of visual representation, from pen and ink to the photographic plate and now to digital and electronic means. The Hubble Space Telescope is only the latest product of this legacy. The images it has returned to Earth are striking testimony to the power of the astronomical telescope to expand our knowledge of the universe.

TELESCOPIC ASTRONOMY

Throughout the 17th and early 18th centuries, astronomers used telescopes to steadily improve maps and tables depicting the positions and motions of stars and planets. But some astronomers wanted to pursue a very different goal. They wanted to penetrate deeper into the heavens. These astronomers looked for ways to increase the light-gathering and magnifying powers of their telescopes. The bigger the lens, the more light was gathered; the longer the focal length, the more images were magnified. Some telescope builders explored ways to use reflecting optics, or mirrors, to collect and focus light. These reflecting telescopes did not depend upon the clarity of the glass, and their mirrors

could be made of a bronzelike alloy called speculum that, when polished, presented a highly reflective surface. But speculum mirrors tended to be brittle, and, most frustrating, they tarnished quickly, requiring a complete repolishing. Both lenses and mirrors were prone to fuzzy images (spherical aberration) and lenses were especially prone to spurious color (chromatic aberration). Telescopes slowly became larger and more refined as artisans succeeded in forging better speculum mirrors, but they generally remained less than six to seven inches in diameter. This situation would change dramatically with the reflectors of William Herschel.

WILLIAM HERSCHEL'S PLAN OF ACTION

During the night of March 13, 1781, a Hanoverian organist and amateur astronomer living in Bath in southern England was standing outside his home carefully surveying the heavens with a seven-inch speculum reflector when he encountered an unusual object: a tiny disk. He recorded it, thinking it might be a comet. Yet when astronomers calculated its orbit, they found that its path was that of a planet, not a comet. William Herschel thus became the first person to discover a

William Herschel's 20-foot reflecting telescope.

major planet, now known as Uranus. This momentous event catapulted Herschel into the limelight.

For his spectacular find, Herschel gained a pension from George III as the King's Astronomer. Now Herschel enjoyed a freedom few astronomers shared. As long as Herschel periodically provided amusement at court by showing what could be seen through his telescopes, he could do anything he wished with his pension and his free time. What he chose to do was build the biggest and most powerful telescopes in the world, increasing his power of pene-trating into space. With such state-of-the-art instruments, he could better map the arrangement of the stars in the visible stellar system as well as probe its limits.

The biggest telescope he built had a 48-inch-diameter mirror. This, however, did not prove to be a success. Herschel was not the first, nor would he be the last, astronomer to push the available technology too hard. Much more capable was his telescope with a mirror 18.5 inches in diameter with a focal length of 20 feet. Herschel also designed an elaborate wooden framework for holding the telescope securely yet leaving it free to move horizontally in short arcs across the southern sky. In this manner Herschel swept the heavens systematically to determine the structure and extent of the stellar system, as well as to hunt for new nebulae and map their distribution and enigmatic forms. These gargantuan tasks would not have been possible without his dedicated associate, his younger sister Caroline, who rapidly became a notable astronomer in her own right. She also drew William's attention to newly found nebulae, helping him map their distribution.[iv] Over the course of his career, William's views on these diffuse objects shifted, but it seems that at the end of his life he viewed some nebulae as indeed truly nebulous, whereas other nebulae he judged to be collections of stars grown misty through their enormous distance.

Herschel's claims were controversial, but he had followers who explored the nature of the nebulae he had catalogued and who were concerned with the structure and dynamics of the stars: devotees of nebular astronomy.[v] William's son John Herschel, among others, embraced the Nebular Hypothesis elucidated by the French mathematician Pierre Simon Laplace, that the sun and its system of planets were formed out of a vast spinning and contracting cloud of ethereal fluids. Characteristics of the theory were linked by many to William's observations. But many other astronomers were not convinced Laplace was right. The decisive issue for opponents and proponents of the Nebular Hypothesis was the true nature of the nebulae, a question bound up with building bigger telescopes, which would allow for better views of the heavens.

ROSSE'S SPIRAL NEBULAE

The Parsons family of Birr Castle in Ireland delighted in constructing novel machines: electrical dynamos, steam engines, suspension bridges, and, most unusual, huge telescopes. William Parsons, the third Earl of Rosse, was fascinated by reflectors. He built a 36-inch-diameter speculum telescope in the late 1830s, and with its aid he tackled the nature of the nebulae, by then a problematic issue, partly because there were as yet no standards for how to use images of nebulae as scientific evidence.[vi]

When the earl and other observers directed the 36-inch telescope at nebulae earlier examined by Herschel, they decided many were star clusters. Herschel's telescopes had shown many nebulae seen by

1.11.95 The Cat's Eye Nebula, shown by Huggins in 1864 to be truly gaseous.

The third Earl of Rosse's depiction of M51, revealing its spiral structure.

earlier observers, like the mid-18th-century observer Charles Messier, to be clusters as well, setting the precedent that larger telescopes would resolve more and more nebulae into stars, hence calling into question the very existence of true nebulae. Starting in 1845, with a monster telescope with a six-foot speculum mirror, dubbed the Leviathan of Parsonstown, Rosse and his associates reckoned they had resolved still more nebulae, including that in Orion, into stars. But Rosse also found a wholly new class of nebula when he turned to Messier's famous 18th-century catalogue and its fifty-first entry, or M51, in the constella-tion of Canes Venatici (the Hunting Dogs), under the cradle of the Big Dipper's curved handle. After several weeks of careful scrutiny, Rosse and his observers and draftsmen recorded and then rendered spiral filaments protruding from the core of the object, suggesting to them that it was shedding material or fleeing through space. Rosse had found a new species of nebula, raising novel questions: What were these spirals, and would even larger telescopes eventually resolve them into stars as well? Or were they evidence of truly nebulous material in space? The next twist in this debate would come from an unexpected direction—the new application of spectroscopy to astro-nomical observation.

A New Way of Looking

On the night of August 29, 1864, William Huggins, observing from his home in London, turned his eight-inch refractor to a small bright planetary nebula in the northern constellation of Draco, the Dragon. Huggins was not measuring the position of the nebula or trying to draw its structure. Instead, he had attached a small spectrum apparatus to his telescope, a device using prisms that took concentrated telescopic light and broke it into a rainbow spectrum so that its chemistry and physical structure might be investigated despite its incredible distance from Earth. Huggins was armed with the latest theory of spectrum analysis established by two German scientists, Gustave Kirchhoff, a physicist, and Robert Bunsen, a chemist, who had shown that a spectroscope could reveal the physical state and chemical composition of an object emitting its own light. One could, indeed, determine what stars are made of.

As Huggins peered into the eyepiece of his spectroscope at the nebula in Draco, known today as the Cat's Eye, at first he saw only darkness instead of the usual rainbow of color, save for a single bright green line. Once he decided that his equipment was not at fault, the laws of spectrum analysis and common laboratory practice told him that this kind of spectrum was produced by a gas, and not by a liquid or solid. Soon Huggins located two other bright, or emission, lines. Even though its bright lines remained unidentified for over 60 years, Huggins had established that the nebula was definitely not a collection of stars; a cluster would instead have exhibited a continuous spectrum like the sun.

Huggins turned his telescope to other centrally condensed nebulae in Cygnus and Taurus that had first been noted by Herschel, and again he found the same type of bright-line spectrum. He next searched out the famous Ring Nebula in Lyra, with the same results.

Huggins also turned his telescope to spiral nebulae. The spectrum of the Great Nebula in Andromeda gave no hint of bright lines, but the Orion Nebula, the one that Rosse and others had declared to be stellar, displayed bright lines, the signatures of a hot gas. Huggins found characteristics of a star only when he trained his spectroscope on the bright blue stars in the Trapezium, at the heart of the Orion Nebula. Otherwise, without exception he found the same

Rendering of the Milky Way from
7,000 light-years above its plane.

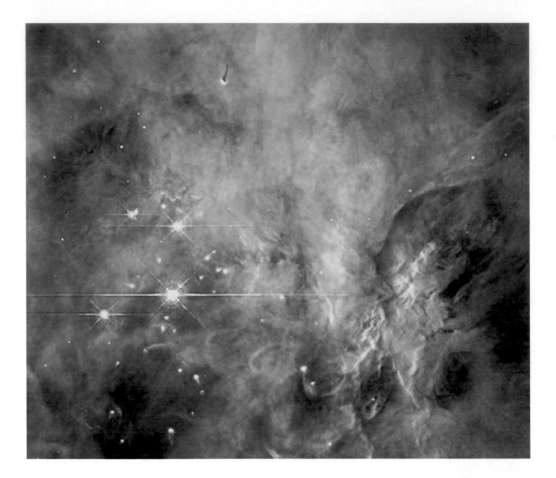

8.24.00 HST image of the center of the Orion Nebula highlighting stars of the Trapezium.

three bright green lines against a black background. When he penned his first report to the Royal Society, Huggins was sure the bright-lined objects were not collections of stars, and concluded with a flourish: "We…find ourselves in the presence of objects possessing a distinct and peculiar plan of structure."[vii]

THE PHOTOGRAPHIC UNIVERSE

Another important tool to explore the stars was developed at the same time. Observers since Herschel's day had wondered if gravity caused nebulae to change shape, and, if so, if this change could be seen by comparing observations of nebulae taken at widely separated times. Although different observers had already drawn differing views of the same nebulae at different times, these might well have been the result of subjective factors. By the late 19th century, many astronomers looked to photography as a way to escape the subjectivity of the observer. As one report put it, through photography "stars should henceforth register themselves" without the intervention or bias an observer would introduce.[viii] Astronomers rejoiced that photography increased reliability, producing a permanent record "untainted by distraction, ill discipline, or bias."[ix] Thus in concert with spectroscopy, photography became a central tool in the study of the physical nature of nebulae.

The powerful combination of photographic recording and large reflecting telescopes forever changed astronomy. Even early photographic emulsions were more sensitive than the eye because photography accumulates light over time, whereas the human eye does not. Glass-backed photographic plates made large-format, wide-field surveys of the deepest realms of the sky possible and provided a permanent record of an object's position or overall character. At more progressive observatories, astronomers were taking steps to replace the refractor with the more versatile reflector, using Leon Foucault's technique of depositing

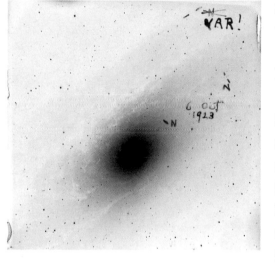

10.6.23 E. Hubble's negative of M31 with his notation identifying a variable star.

silver on glass. That meant they could dispense with heavy speculum metal mirrors. The Lick Observatory in California installed a 36-inch reflector purchased from Edward Crossley in England, and with it James Keeler began examining the enigmatic spiral nebulae with long-exposure photography. Keeler's surveys convinced him that upward of 700,000 spirals existed within in the range of his telescope.

A decade after Keeler's death in 1900, Vesto Melvin Slipher, working at Lowell Observatory in Arizona, improved the capabilities of his spectrograph, and with its aid he secured several exceptional finds. Most baffling were the shifts he found in the spectral lines of many spiral nebulae. Slipher interpreted these displacements, using the Doppler effect, to mean that many of these objects were traveling at great speeds, far faster than any star. Indeed, some of Slipher's nebulae were traveling over a thousand kilometers per second, unbelievably fast for any terrestrial mind to comprehend.

Edwin Powell Hubble

Namesake of the Space Telescope

Edwin Powell Hubble was born in 1889 in Missouri. He studied at the University of Chicago and received his Ph.D. at Yerkes Observatory in 1917 for a meticulous reconnaissance of the forms of faint spiral nebulae. After wartime service, Hubble was hired by George Ellery Hale in 1919 and immediately put to work at Mount Wilson.

Hubble was relatively junior for an astronomer when he made the momentous discovery of the distances to spirals in 1924. His employment at Mount Wilson was important, but his name was hardly known. As a result, Hubble had two major hurdles to overcome. The first was proving the remoteness of the spiral nebulae using the Cepheid relation. And the second, more subtle but nonetheless equally critical, was the fact that a senior Mount Wilson astronomer, Adriaan van Maanen, had been collecting evidence that showed the spirals to be in rapid rotation. Matching photographs of these spirals taken over several years of

time, he believed he had detected actual shifts in position of the arms. Taken along with the spectroscopic determinations of their velocities of rotation, Van Maanen and others were able to estimate the distances to these nebulae, and found them relatively close by. Van Maanen was only five years older than Hubble, but was almost ten years advanced over him professionally. Upon his subsequent arrival at Mount Wilson, he had quickly established himself as an expert in positional photographic astronomy. Hubble knew that his findings would make Van Maanen a formidable adversary. He spent much of 1924 redoubling his efforts to present a solid case for the remoteness of the spirals. Hubble's confidence to publish his results was strengthened when influential astronomer Henry Norris Russell urged him to send a report to the national meeting of the American Association for the Advancement of Science.

Hubble was unable to attend the meeting, so Russell offered to read it

Margaret Bourke-White study of Edwin Hubble at the 100-inch Hooker telescope.

for him. The astronomers in attendance voted to send it to the AAAS Awards Committee; it shared the prize that year.

From this auspicious beginning, Hubble continued to expand knowledge of the universe, opening up new doors in astronomy. With collaborators Richard Tolman and Milton Humason in the 1930s, Hubble addressed fundamental questions about the nature and structure of the universe, and tirelessly continued to observe its depths.

Slipher's puzzling observations, as well as Keeler's astounding conclusions, posed mysteries to scientists that could not be solved until the advent of bigger telescopes—and a new type of astronomical practice.

THE INFRASTRUCTURE OF ASTRONOMICAL DISCOVERY

George Ellery Hale was the quintessential American booster for building both big telescopes and big scientific institutions. A child of industrial and corporate America, raised in Chicago in the 1870s and '80s, he was a genius at persuading newly rich American patrons to construct big telescopes. From the 1890s to the 1930s, he built the largest telescopes in the world, four times: the 40-inch Yerkes refractor, the 60-inch and 100-inch reflectors at Mount Wilson, and finally the 200-inch Hale telescope at Palomar.[x] A land that could make great battleships and bridges, Hale once mused, could always build a bigger telescope.

With the support of the recently established Carnegie Institution of Washington, a 60-inch telescope was completed in 1908 that combined three telescopes in one, using an interchangeable set of smaller "secondary mirrors" to reflect and concentrate the light in different ways. A single flat secondary mirror pro-

vided the simplest and shortest focal length, photographically the fastest system. Convex secondary mirrors created optical systems with much longer effective focal lengths. One of the convex mirror systems was very similar to a design the Hubble Space Telescope would later employ, called a Cassegrain telescope.

With the 60-inch telescope, Hale's staff was soon taking high-resolution photographs of star clusters and spiral nebulae. Between the years 1914 and 1919, Harlow Shapley used the giant telescope to photograph globular clusters. These great balls of stars were scattered through the sky but seemed to be concentrated around the brightest region in the Milky Way, in the direction of the constellation of Sagittarius. Shapley hunted in them for a class of variable star called a Cepheid (after the prototype found in the constellation of Cepheus). The variation in a Cepheid's light output was an indicator of how bright the star was. With its brightness known, astronomers could calculate its distance. With the assistance of the Cepheids, Shapley measured rough distances to dozens of globular clusters.

In the 1910s, nearly everyone believed that the sun was very close to the center of the Milky Way.

Shapley, however, arrived at a very different conclusion. He reckoned that the globular clusters form a spherical system, centered on a point in the Milky Way somewhere between 60,000 and 100,000 light-years away from the sun. If this was true, the Milky Way was a disk of stars 300,000 light-years in extent, far larger than anyone had previously thought. And the sun was placed tens of thousands of light-years from its center.

Even as Shapley puzzled over the globular clusters, Hale was intent on building a much bigger telescope. Supported by private and corporate philanthropy, Hale enlisted teams of experts ranging from glassmakers and optical specialists to battleship manufacturers to build a 100-inch telescope, its mirror, mounting, and surrounding dome. Finally, the giant telescope was ready. In November 1917, Hale, with his senior staff astronomer Walter Sydney Adams and the British poet Alfred Noyes, peered through the telescope. Of the important moment Noyes later wrote: "Where was the gambler that would stake so much—Time, patience, treasure on a single throw?" [xi]

Hale was indeed a gambler. But within a very short time, his gamble paid off. By 1920, true to his philoso-phy of providing a flexible test bench to test physicist's dreams, the telescope had been equipped with a device that actually measured the angular diameters of red giant stars, far beyond the normal resolution limits of the Earth's atmosphere. And bigger revelations were soon to come.

The completed telescope was the most powerful ever built, and Hale hired a young astronomer named Edwin Hubble to be one of those to use it.

THE PUZZLE OF THE SPIRAL NEBULAE IS RESOLVED

In the fall of 1923, Hubble was allotted many nights to observe with the great 100-inch reflector on Mount Wilson. Starting in late September, Hubble photographed the Andromeda Nebula (M31) repeatedly from the shortest and fastest focus of the 100-inch telescope. This required him to sit on a platform high off the floor and carefully guide the photographic plate holder with precision screws for hours. By the end of the year and numerous nights of observing, Hubble had collected and examined many sets of photographs. On one particular plate, taken in early October, he found what he at first thought was a nova, a star that brightened rapidly and violently to a luminescence far

3.7.01 The jumbled chaos of newly formed stars following an encounter of two galaxies.

outshining all other stars. But when he examined another photograph taken a bit later in the sequence, he found that, unlike a nova, the star quickly dimmed. Plates taken later showed it brightening again. From the pattern of its light variations, Hubble decided, here was a Cepheid variable. He quickly detected a second example.

Those two variables enabled him to estimate the distance to M31, around one million light-years. With other finds rapidly following in the wake of the first two, Hubble swiftly settled the question of the nature of the spirals: They were indeed island universes well beyond the borders of the Milky Way and comparable to it in size. The universe was not one of stars, but one of galaxies. [xii]

With the most powerful telescope in the world, sitting in the mile-high dark mountain air above Los Angeles, Hubble was master of the extragalactic realm. He turned his attention to the strange spectral shifts Slipher had found in the spectral lines of many spiral nebulae. He gradually built up a means of determining distances to ever fainter—so presumably more distant—spirals that allowed him to calculate how the spectral shifts of galaxies changed with distance. By 1929, Hubble was confident he knew the answer: There was a definite, though yet rough, correlation between the distance of the galaxy and the shift of its spectrum. Double the distance, double the shift toward the red end of the spectrum.

In the 1930s Hubble and Milton Humason continually refined this so-called red shift relation. Others were quick to offer interpretations of Hubble's observations. The red shifts revealed that the universe had no center to its expansion. The rate of expansion, later to be known as the Hubble Constant, became the Holy Grail of cosmology. Hubble first calculated the rate of expansion term at five hundred kilometers per second for every megaparsec (one million parsecs—a parsec is approximately 3.26 light-years) one traveled into space.

Hubble also concluded that Keeler's estimate of 700,000 visible spirals (or galaxies, as almost all astronomers now judged them to be) had to be revised upward into the millions and soon hundreds of millions. By 1936, Mount Wilson observers had detected velocities as high as 40,000 kilometers per second for a faint cluster of galaxies in Ursa Major. There seemed to be no end to the universe, but Hubble wondered what the recessional velocities really meant. Was the universe truly expanding, or was the apparent expansion an artifact of some effect of space and time not yet understood?

Despite Hubble's conservatism, astronomers have built upon his work to create a master plan. In the words of one of his successors at Mount Wilson, galaxies exist and they are moving away from one another. Adding to the plan became the call to arms for most of the rest of the century: Refine the observed rate of expansion, determine if it is linear or varies with time, and, most of all, explore the nature and extent of the vast and apparently boundless system of galaxies that constitutes our universe.

The Early Days

This chapter's gallery showcases a variety of objects that reveal various stages in the lives of stars.

3.1.01 Profile of a galaxy, revealing lanes of absorption by dust and gas.

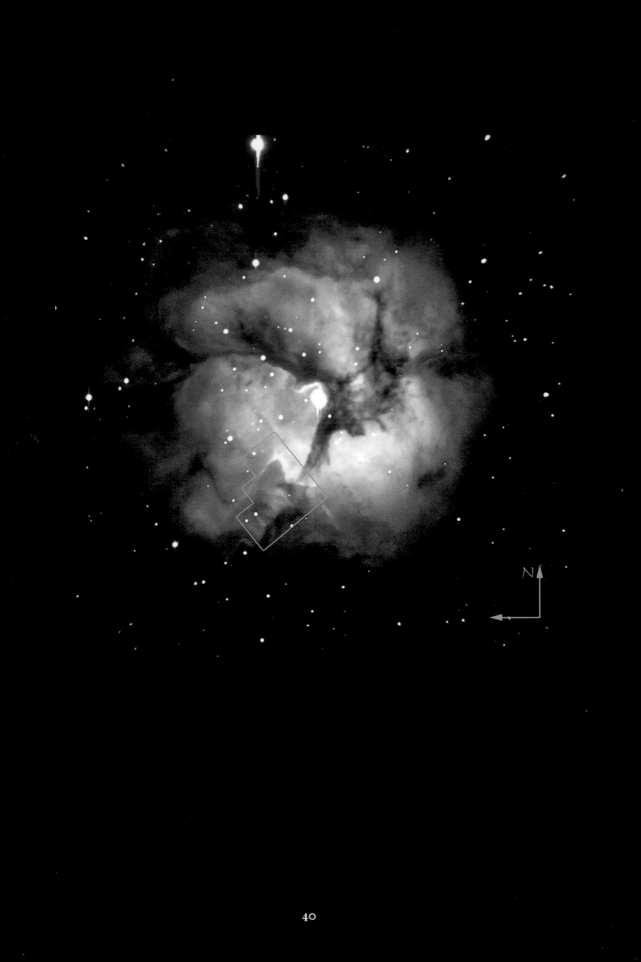

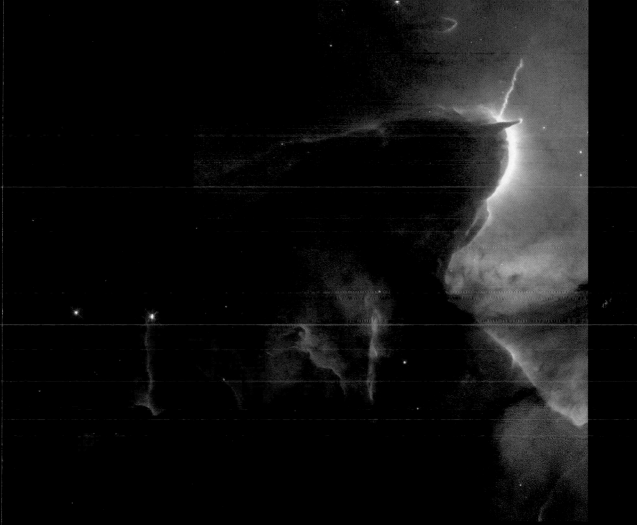

LEFT: 11.9.99 Hale telescope image of the Trifid Nebula in Sagittarius.

ABOVE: 11.9.99 HST image of portion of Trifid revealing signatures of embryonic stars.

5.14.98 The central portion of Centaurus A showing regions of intense star formation.

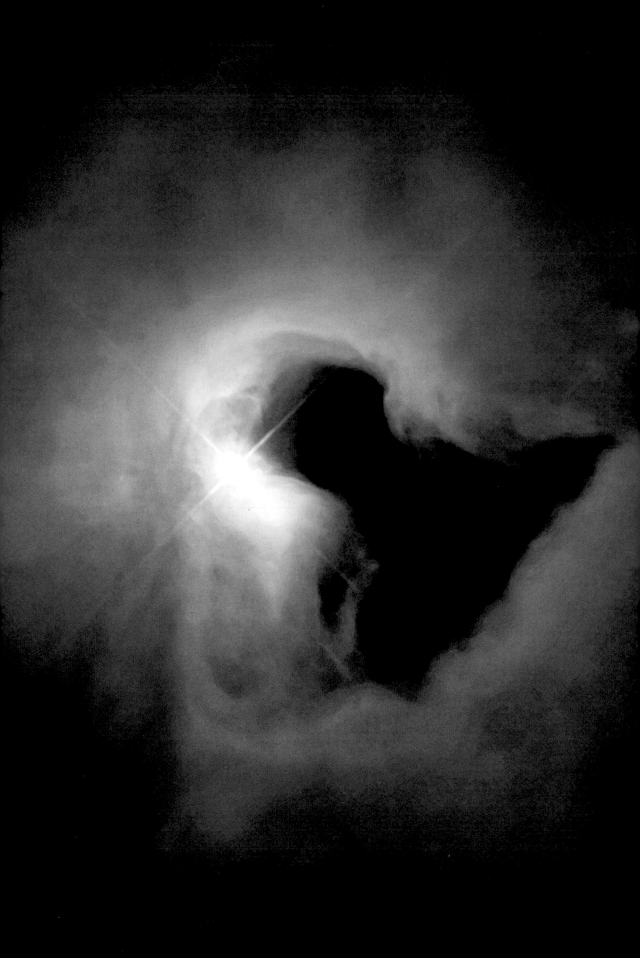

LEFT: **3.2.00** Reflection nebula in Orion illuminated by a bright young star.

ABOVE: **8.1.02** Gomez's Hamburger, a star ejecting a shell of gas and dust in Sagittarius.

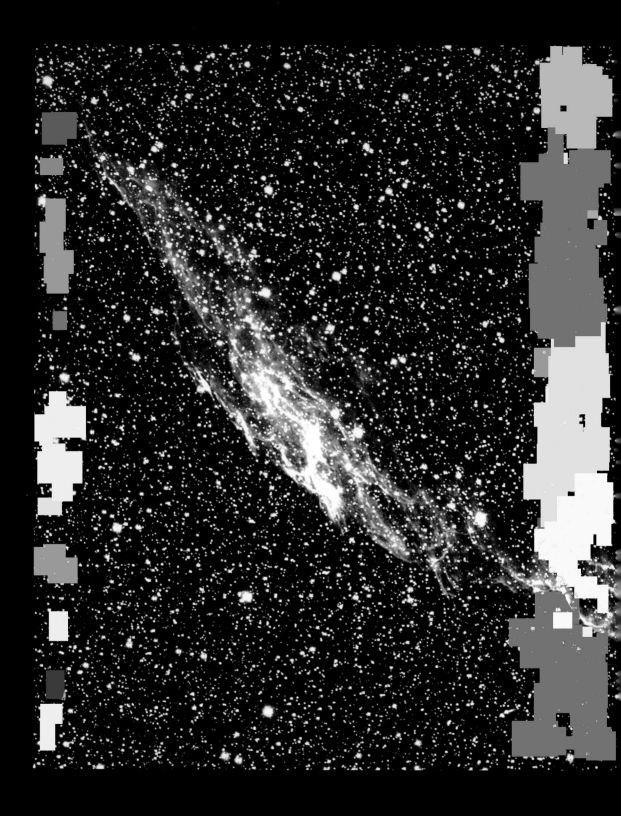

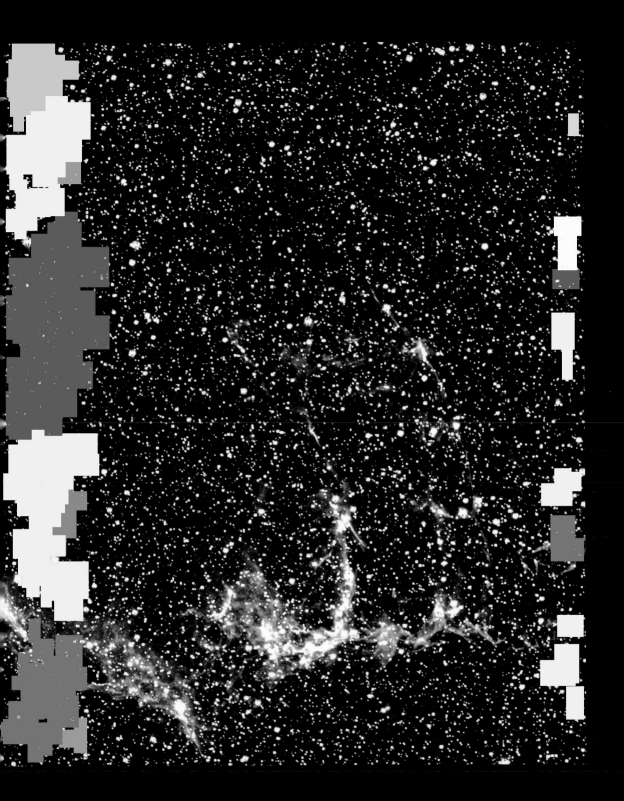

7.15.96 Ground-based image of the Veil
Nebula, a supernova remnant.

47

e Focus. f 3.3
rain . f 16.
DE . f 30

APPROXIMATE SCALE

R. W. PORTER '3

The 200-inch Hale Telescope at Palomar, state-of-the-art when Spitzer first voiced his dreams

The Early Days

2

Acting on a Dream

THE HUBBLE SPACE TELESCOPE (HST) WAS LOFTED INTO ORBIT ON APRIL 24, 1990, ABOARD THE SPACE SHUTTLE *DISCOVERY*. ITS EVOLUTION UP TO THAT HISTORIC DAY WAS REMARKABLE IN MANY WAYS. WELL OVER 20 YEARS OF HIGHLY DETAILED PLANNING HAD INVOLVED THOUSANDS OF ASTRONOMERS, ENGINEERS, AND MANAGERS, AS WELL AS THE EFFORTS OF HUNDREDS of universities and industrial contractors across North America and Europe. The 12-ton telescope, the most complex multi-instrumented spacecraft ever launched, embodied the primary aim of astronomers: to penetrate deeper into space. Even so, the building of the HST had a checkered history. On several occasions it seemed that the project might be canceled. At times both the technical and political challenges of building and flying such a sophisticated telescope appeared almost insuperable. Yet all of these hurdles would be overcome.

One of the astronomers on hand to watch the Space Shuttle rise into the blue Florida sky at Cape Canaveral that April morning was Lyman Spitzer, Jr. In a 1946 report, "Astronomical Advantages of an Extra-Terrestrial Observatory," he had pondered what sorts of astronomical observations could be made from an artificial satellite with a large telescope. As had George Ellery Hale when promoting his own schemes for giant telescopes, advocates like Spitzer of a big telescope in space discussed particular scientific problems that such an instrument could address. But they also emphasized the extra capabilities it would offer. Such a leap in observing power was sure, they contended, to result in original and profound findings.

A large space observatory, Spitzer conceded, was nevertheless years away. As he was writing in 1946, when no man-made object had ever been lofted into Earth orbit, he could hardly claim more. Although in the late 1940s a few researchers were experimenting with what observations could be made from rockets sent above the atmosphere, or with sensitive radio equipment on the ground, these techniques were not as yet major research areas in astronomy. This does not mean that astronomers failed to recognize the exceptional potential of observing the universe without having to worry about the obscuring layers of the Earth's atmosphere. As one influential astronomer quipped in 1933, when astronomers died, they should be "permitted to go, instruments and all, and set up an observatory on the moon." From the moon's airless surface an astronomer would not have to contend with the vagaries of the weather and location.[i]

Spitzer, as would any astronomer, keenly appreciated how the Earth's atmosphere weakens, distorts and blocks radiation from space. We live under a turbulent ocean of air, constantly in motion, consisting of dissimilar regions. Astronomers think of this ocean as made up of packets called cells that range in size from a few centimeters on up. A beam of light passing through the atmosphere encounters cells of differing temperature and density, and the cells are always moving. This causes the beam to be distorted and bent, and consequently, the eye will see the beam of light flicker or, as it is commonly described, twinkle. A large telescope gathers many of these twinkles together so the image it forms is composed of lots of small images that produce a blurred image far larger than any one twinkle during a time exposure.

The upper atmosphere also glows faintly and blocks virtually all ultraviolet rays, x-rays, and gamma rays, and most infrared rays as well. The latter condition is good for our health, but keeps much critical information from an Earth-based telescope's observation. Although astronomers placed their telescopes on high mountains or carried them aloft in aircraft and under balloons, they could only dream about a large telescope in space, free of all the restrictions of the atmosphere. Throughout the 1950s, the reality

Artist's depiction of a planetary system forming around star Beta Pictoris.

seemed decades away. But in the 1960s, changes in space travel began to accelerate much more quickly than anticipated.[ii]

HOW LARGE A TELESCOPE?

The launch of Sputnik 1 in 1957 set in motion the space race between the United States and the Soviet Union. Money for space endeavors now became available as never before, and vast sums were devoted not only to fly astronauts in contests for national prestige, but also to advance space science, including space astronomy. Spitzer's dream of 1946 transmuted rapidly from the world of science fiction into the realm of the possible. During the 1960s, the newly formed space agency National Aeronautics and Space Administration (NASA) promoted a range of space astronomy activities. One was planning for what was generally termed by the end of the decade the Large Space Telescope, or "LST" for short.

Astronomers, engineers, and industrial contractors teamed up with NASA staff, particularly at the Marshall Space Flight Center in Alabama and the Goddard Space Flight Center in Maryland, to study what the LST might look like and do. They all agreed on one issue quickly: the proposed size of the primary mirror. An earlier study at NASA's Langley Research Center showed that a primary mirror of 120 inches diameter was the largest that could be built for optimum performance and still be transported into space by one of the Saturn rockets, the most famous of which was the Saturn V, which powered U.S. astronauts to the moon. The size stuck. Astronomers in general argued that the bigger the LST mirror, the bigger the likely scientific payoff. But these key astronomers realized that if they were too ambitious, it might lessen the chances the LST would ever be built. A 120-inch mirror offered the prospect of extremely exciting scientific gains but was not so large that it would make LST prohibitively expensive or too demanding technically. It would not be as large as the largest ground-based telescope, such as Palomar's 200-inch in southern California, or the four-meter-class telescopes that were then being planned, but it would be large enough to provide views from beyond the atmosphere that would more than compensate for its relatively small size. The aim was for the LST to be so well engineered, capable of being pointed so precisely and holding its gaze so accurately, that by escaping the atmosphere its images would be practically perfect. The LST soon came under fire. Congress worried about the cost of the project, expected by 1974 to be several hundred million dollars, and the most obvious way to save money was to reduce the size of the primary mirror and the spacecraft. By 1975, with the threat of cancellation hanging in the air, planners looked for ways to cut

Lyman Spitzer, Jr., a key figure in HST's history, in the late 1940s.

NASA's depiction of a downsized Space Telescope deployed by shuttle.

costs, reducing the 120-inch mirror (3 meters) either to 2.4 or even to 1.8 meters. NASA engineers' estimates showed what savings would in fact be gained. The reduction to 2.4 meters saved $61 million, but to reduce all the way to 1.8 meters, which astronomers objected to strenuously, saved only another $14 million. Thus NASA settled on 2.4 meters, at a projected cost of $273 million.

Costs remained high because the telescope's basic requirements for pointing accuracy, stability, power, and data handling were still highly demanding. Astronomers on scientific review boards generally accepted the 2.4-meter compromise, so the project was redefined with that size mirror by the end of 1975. A 2.4-meter mirror in space could still examine fine details of astronomical objects in the sky far better than any optical mirror from the ground. Its resolving power would be equivalent to being able to distinguish the left and right headlights of a car in California from a telescope as far away as New York, or features less than 1/30,000th the size of the full moon. This was at least a tenfold increase over the atmospheric limit. Reflecting the reduction in size, the Large Space Telescope became the more modestly named Space

An Eye in Space

The Initial Complement of Instruments

When the Hubble Space Telescope was launched in April 1990, it carried aboard a complement of five dedicated scientific instruments. Four were axial instruments; that is, they were inserted into the telescope parallel to its main optical axis. These instruments were all roughly three by three by seven feet in size and weighed several hundred pounds. Wide Field/Planetary Camera 1 was a radial instrument; it was inserted radially into the telescope and was essentially wedge shaped. All of the instruments were designed to be modular so astronauts could repair or replace each separately in space. To hold the instruments securely in place to very precise tolerances yet still easily allow an astronaut in a bulky spacesuit access to them, 27 special latches had to be fabricated. The latches were extensively tested in the gargantuan water tank at the Marshall Space Flight Center that contained a full-scale mock-up of the HST.

Faint Object Camera

Built by the European Space Agency, the Faint Object Camera was designed to make full use of the HST's ability to resolve fine detail in astronomical images but had a small field of view. It contained two complete and independent cameras, each with its own optical path and detector system, one working at a camera speed of f/96, the other at f/48.

Faint Object Spectrograph

Intended to observe very faint objects in both visible and near ultraviolet regions of the spectrum, it employed a type of one-dimensional light detector known as a digicon. Digicons convert light falling on them into a stream of electrons, which are then focused on a string of silicon diodes. The diodes transform the different levels of light intensity into electrical signals.

Goddard High Resolution Spectrograph

Designed to observe in the ultraviolet region of the spectrum, it was fash-

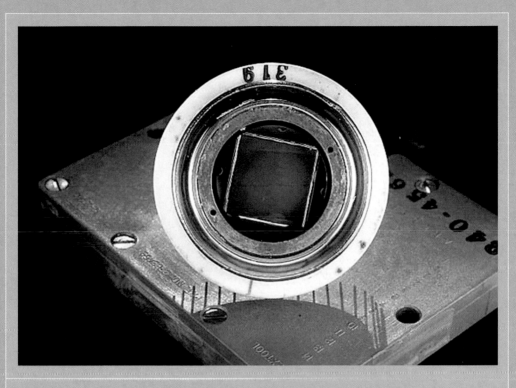

The tiny CCD, the silver-gray square, drove the redesign
of the Wide Field/Planetary Camera.

ioned to analyze the light of relatively bright objects in high resolution. It also employed digicons as light detectors.

HIGH SPEED PHOTOMETER

Mechanically the simplest of all the instruments, it contained no moving parts and was designed principally to measure rapid changes in brightness of an astronomical object over times as small as ten microseconds. HST's pointing system was used to direct light precisely into a combination of tiny filters and apertures in this instrument.

WIDE FIELD/PLANETARY CAMERA

This instrument carried eight CCD postage stamp-size detectors, arrays of 800x800 pixels that acted like very sensitive rewritable electronic film. It could be operated in two ways, each using four of the CCDs. To image extended faint objects such as galaxies and diffuse nebulae, the Wide Field Camera was used. The Planetary Camera was designed to image small bright objects such as planets using a narrower and higher-powered field of view.

There was also in effect a sixth instrument: The telescope's three Fine Guidance Sensors could also act as accurate position measuring devices.

Telescope. In 1983 it was renamed the Hubble Space Telescope.

DESIGNING THE MIRROR AND THE OPTICAL TELESCOPE

By the late 1960s, astronomers were very familiar with a compact and efficient form of a reflecting telescope with outstanding optical qualities. In this arrangement, light from an astronomical object travels down the telescope tube and reflects off the primary mirror (shaped like a hyperboloid) back up the tube to the secondary mirror (also a hyperboloid). Instead of being directed out of the side of the tube, as in the case of a typical amateur telescope, or like the 100-inch at Mount Wilson, the light is bounced back toward the primary mirror and through a hole in its center, where it can be analyzed. Known as a Ritchey-Chrétien after its designers, George Willis Ritchey and Henri Chrétien, who first suggested it at Mount Wilson around 1910, it provides wide field views of exceptional fidelity and has been used on some of the world's finest ground-based telescopes like the Mayall four-meter reflector at Kitt Peak. The optical system is folded and the scientific instruments are located behind the primary mirror, a big advantage in balancing a bulky and massive space observatory.

For this reason, there was little controversy on the telescope's overall optical configuration. The Ritchey-Chrétien was widely judged as the obvious choice.

Astronomers and engineers did debate how the mirror should be designed and built. The mirror had first to survive the extreme rigors of launch and next of exposure to space itself. It was critical that the mirror not become distorted no matter what conditions it encounterd. Otherwise it would not produce high-quality images. One possible mirror material NASA considered was fused silica (quartz), which was well known to be stable over long periods. Also examined was a newer sort of glass, ultra low expansion (ULE) glass, which did not expand or contract as much as quartz when it was heated up or cooled down. As a test, the Itek Corporation created a special lightweight egg crate mirror of ULE glass, and polished and tested it. The egg crate design, consisting of interleaved vertical slats between two circular plates, was far lighter than a solid mirror. This was an attractive feature, since engineers were trying to save weight wherever possible in designing the spacecraft. The U.S. Air Force had also already developed such mirrors for spy satellites, so the technology was deemed acceptable.

4.24.90 HST in acoustic test chamber at Lockheed, Sunnyvale, California, in the late 1980s

Optician poses with 200-inch Palomar mirror blank and a Galilean telescope.

LOCATION, LOCATION, LOCATION

In the 1960s, the Marshall Space Flight Center and some of NASA's industrial contractors studied a space station-based scenario that would permit astronauts to change photographic plates in orbit. Most astronomers, however, much preferred an orbiting observatory that could be operated remotely, or an observatory that astronauts might visit to effect repairs or replace components.

In the early 1970s NASA had its own top priority: to develop a space shuttle fleet. Because the prospect of a permanent space station looked ever more remote by then, NASA rearranged its plans concerning the Space Telescope. Previously the telescope had been earmarked for launch aboard a Titan III rocket, but after President Nixon approved the shuttle in 1972, NASA mated the telescope to it both as its launch vehicle as well as the means to carry astronauts to service it. The upper limits to the telescope's size and orbit were therefore set by what could be carried in the shuttle's payload bay and then serviced by the shuttle. Instead of building the Space Telescope and flying it until it failed, NASA had planned an extended lifetime in orbit, 15 years at least. By touting the use of the shuttle to effect repairs in orbit, NASA also claimed that the telescope would need less testing on the ground than the usual spacecraft, a point that would help reduce overall costs. This, as it turned out, was at best a dubious proposition.

A crucial question confronting scientists was how to return the telescope's scientific data to Earth. By the early 1970s, the United States had extensive experience in flying spy satellites that had been designed to secure information on foreign countries, principally the Soviet Union, from space. Some of these satellites took photographs and then ejected the film in specially designed return capsules to be snagged by aircraft as they floated down through the atmosphere on parachutes. Although several of the contractors working on the telescope were the same ones working on spy satellites, such a scheme was never seriously discussed for the Large

10.4.90 WF/PC I images an outer portion of the Orion Nebula.

Space Telescope. Instead, planners assumed that the data amassed by the telescope's scientific instruments would need to be transmitted back to Earth electronically.

THE INSTRUMENTS OF PERCEPTION

Hale's concept of a multifunctional infrastructure for astronomy would bear full fruit in the HST. From early in the planning stages, it was designed to be a long-lived "observatory"— not a mere telescope. By 1975, plans for the telescope included instruments with a wide array of scientific capabilities: cameras, photometers, spectrographs, and an astrometer—an instrument that measures the positions of astronomical objects. In the ensuing debate over specific designs, two were chosen as core instruments, the Wide Field Camera (WFC) and Faint Object Spectrograph (FOS). They were deemed core because they were widely judged to be critical to the telescope's goals.

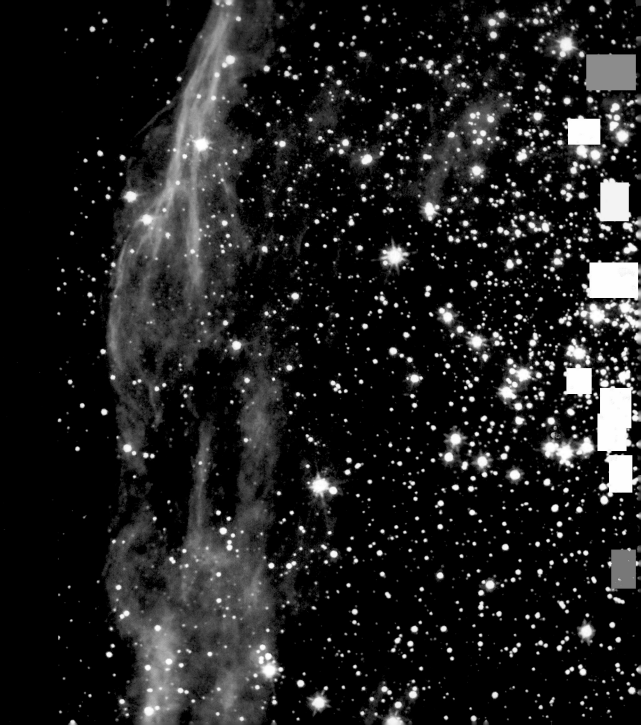

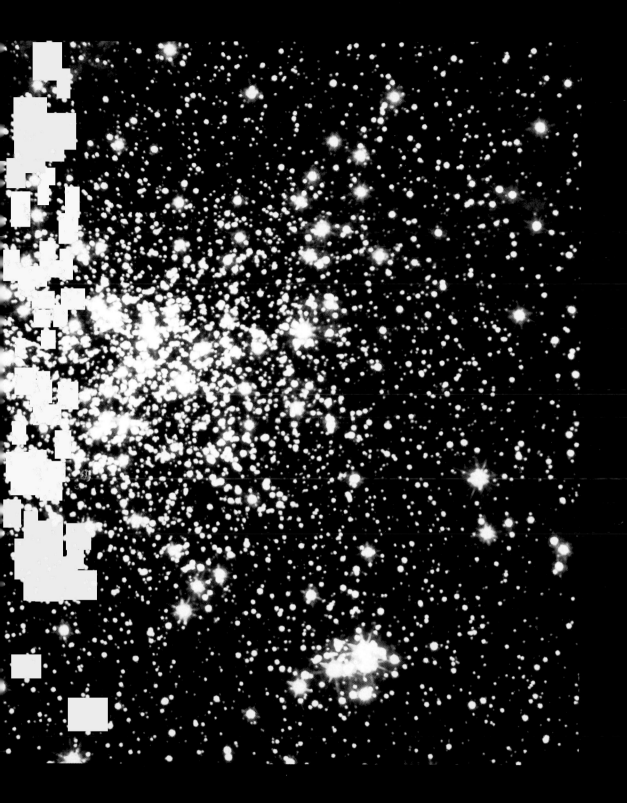

7.10.01 High resolution image of double cluster of stars, after 1993 repair.

Many astronomers, however, judged the Wide Field Camera to be the key instrument. Because the telescope's principal scientific goals were cosmological, focused on the faintest and most distant objects in the universe, astronomers wanted a camera with a relatively large field of view. The desired size was three-by-three arc minutes, which would capture in one frame distant clusters of galaxies. Although this field was barely 1/100th the size of the full moon in the sky, it was comparatively large by most telescopic standards.

Initial designs of the Wide Field Camera employed a sensitive television tube called a secondary electron conduction (SEC for short) vidicon as its light detector. Such tubes were already used for space missions because they worked well in the ultraviolet range, and the images they formed could be readily transmitted back to Earth. But there were other factors weighing against them. SEC vidicons consumed a lot of power, were bulky, and employed fragile vacuum tubes. They were also not sensitive to red colors and were not well adapted for bright objects such as planets. In the mid-1970s, these limitations worried planetary astronomers, who foresaw many uses for the telescope to observe the planets, moons, comets, and asteroids within our solar system. They lobbied for a new form of solid-state detector, called a charge coupled device, or CCD. An image from the telescope would be guided by the camera's optics onto the CCD chip and produce a sort of electronic photograph. The chip would create a facsimile of the scene, broken down into a large number of discrete picture elements, or pixels. The information from each of these pixels would be handed, bucket-brigade fashion, to another area of the chip, stored with an identification tag, and then passed on to Earth.

The CCDs of the early 1970s, even though small and relatively simple chips, offered the prospect of major advantages for space operations over vacuum tubes. In fact, these chips appeared so attractive in the long term that they rapidly drew the attention of various companies and the U.S. military, as well as the staff of the Jet Propulsion Laboratory (JPL) in California, the key center for NASA's program of planetary spacecraft.

Yet the CCDs had drawbacks of their own. They were a new and in many respects immature technology, and if applied to an astronomical

Telescope Aberrations

Spurious Additions to Telescopic Observation

There is no such thing as a perfect telescope; they all have aberrations. Some are due to imperfections in the lenses; others are caused by improper alignment; sometimes a lens or mirror surface has been incorrectly shaped. Inherent imperfections exist in even the simplest designs of lens and mirror telescopes.

In Galileo's day, astronomers rapidly realized that spurious rings of color seen in the telescope were not caused by the eye. By the end of the 17th century, they had realized that the colors did not come from the glass in the lenses. Anyone who has looked through a prism, or merely a jagged piece of glass at a bright white light, will notice tiny rainbows of color. These colors, as Isaac Newton taught, are inherent in the light beam itself. The glass merely spreads the white light apart into its constituent colors. No one before the 18th century was able to devise a collection of lenses to magnify objects that produced images without "chromatic aberration."

Eventually, astronomers found that lenses could produce virtually color-free images if the curvature of the lens was decreased. As a result, telescopes grew longer and longer, some reaching well over 130 feet, becoming extremely unwieldy to operate. In reflecting telescopes, optics are free of chromatic aberration. Newton knew this, and proposed using reflecting telescopes instead of lenses to achieve color-free images.

A spherical lens or mirror will not bring all rays to a single focus, but a parabolic mirror will; its inner portions have a slightly deeper curve. John Hadley was the first to successfully parabolize a Newtonian reflector in 1721, making them equal if not superior to long refracting telescopes.

Throughout the late 17th century and into the 19th, refractors dominated. When larger and more stable mirrors made of glass emerged, and the problems astronomers attacked changed in the 20th century, reflectors came to the fore and remain common astronomical practice today.

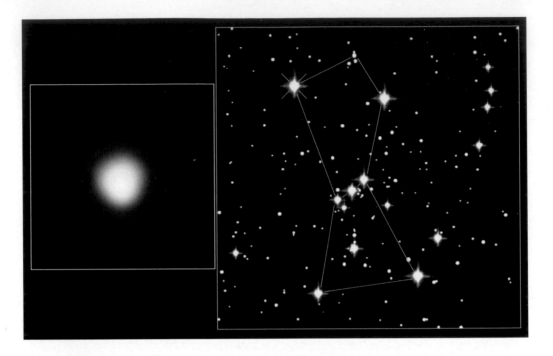

12.10.96 Faint Object Camera images red supergiant star Betelgeuse, in the shoulder of Orion.

camera, they would cover only a very small field of view, certainly much smaller than a vidicon. To overcome the field limitation, a team from Caltech and the Jet Propulsion Laboratory, led by Caltech astronomer James Westphal, came up with an ingenious solution. Responding to a NASA competition in 1977 to decide who would build the instruments to fly aboard the HST, they argued that the best way to increase the field of view was to combine four CCDs together, thereby turning them in effect into one big chip. In their scheme, light captured by the telescope and directed toward the camera would fall first on a pyramid with four

mirrored faces that would split the light into four beams, each illuminating one of the four CCDs. In addition, the pyramid itself could rotate to two different positions, illuminating two sets of four CCDs, and thereby cleverly combine a Planetary Camera and a Wide Field Camera in the same instrument, producing the Wide Field/Planetary Camera. To enhance the CCDs' ultraviolet response, the winning Caltech-JPL team also proposed to coat them with an organic phosphor sensitive to ultraviolet light.

In the same competition, NASA selected three other instruments to analyze the light collected by the telescope to add to the Faint Object Camera to be provided by the European Space Agency (ESA), which was now a junior partner on the Space Telescope project. Roughly the size of a phone booth, weighing several hundred pounds, each instrument package would be positioned behind the telescope's primary mirror. These three instruments were: a High Speed Photometer (to measure extremely rapid variations in brightness of astronomical objects) and also two spectrographs—a Faint Object Spectrograph and a High Resolution Spectrograph. These would separate the light from various objects into its constituent wavelengths and investigate the result. A sixth instrument, the telescope's Fine Guidance Sensors, would determine astronomical positions with extreme accuracy.

HOLDING STEADY

One absolutely crucial element of the telescope's performance would be its ability to acquire its astronom-

HST's primary mirror after Corning's blank had been ground, polished, and coated.

ical targets and then to stay locked on them for perhaps many hours so that the incredibly faint light from far-off objects could slowly accumulate on the detectors of the various scientific instruments. When astronomers and engineers worked out the pointing requirements, they found the telescope had to be stabilized to an unprecedented 0.012 arc second (by comparison, the full moon has an angular size of about 1,800 arc seconds). This is equivalent to directing a laser in New York toward a dime in Washington, D.C., and keeping the laser trained on the dime. Since the spacecraft would be orbiting the Earth at 17,500 miles per hour, this was a hugely demanding task. Compounding the difficulty was jitter, or tiny shaking motions of the telescope due to the operation of various mechanisms it carried.

Engineers examined two possible approaches to pointing the telescope. In the first, minor adjustments in position could be made by shifting

4.1.92 WF/PC I images a 200,000°C white dwarf star in the center of a planetary nebula.

the body of the spacecraft itself. In the second, very small changes in the position of the telescope's secondary mirror could reposition the images. Finally, the former option was chosen. To provide direction to the spacecraft on which way to point, light from a star would fall on a Fine Guidance Sensor. The sensor would measure the position of the guide star, and the measurement would then be converted into an electrical signal. This would in turn inform the support system module to correct the pointing direction. Using two of the three Fine Guidance Sensors would keep the telescope locked on its astronomical targets.

A Spacecraft Takes Shape

From the early 1970s on, the basic design of the telescope was divided into three parts: (1) the Optical Telescope Assembly (OTA), (2) the Support Systems Module (SSM), and (3) the Scientific Instruments (SIs). The Optical Telescope Assembly would be a folded type of reflecting telescope, at the heart of which was the primary mirror. The Scientific Instruments would be positioned behind the primary mirror and analyze light in a variety of ways. The Support Systems Module would manage the housekeeping functions of the spacecraft. Its tasks included receiving the output of the solar panels and routing the energy to the instruments and systems for heating, pointing, and moving the telescope to new astronomical targets. It also would house systems for data handling and radio communication. Designed to act in many ways as the spacecraft's brains, it would monitor and regulate the spacecraft functions supporting the OTA and SIs. After the decision was made in 1975 to pursue a 2.4-meter telescope, the Support Systems Module was repositioned to wrap around the optical system of the telescope and its instruments like a life preserver, instead of being stationed behind the primary mirror, as it had been envisioned earlier.

Downsizing to a 2.4-meter mirror may have disappointed the astronomers, but the new size boosted the confidence of engineers that they could indeed build the observatory they were designing. Pointing the telescope was now simpler. The difference in weight between the 3-meter and 2.4-meter versions of the telescope was 8,000 pounds, meaning it would better fit with the weight and size of payload that the space shuttle could carry to orbit.

By 1977, the main design features of the Space Telescope had been set. Approximately 45 by 15 feet in size, or about that of a school bus, it would be a 2.4-meter Ritchey-Chrétien reflector. The primary mirror would be made of ultra low expansion glass in a lightweight lattice array. It would carry six scientific instruments. After a three-year battle to secure funding, the White House and Congress at last gave approval to NASA and its industrial contractors to start building the telescope, and the European Space Agency had agreed to join with NASA to provide a scientific instrument as well as the solar arrays to power the spacecraft. Many compromises had been made, but on the drawing board was what promised to be a superb observatory. At last work could begin on its construction, with astronomers eagerly anticipating the 1983 launch date.

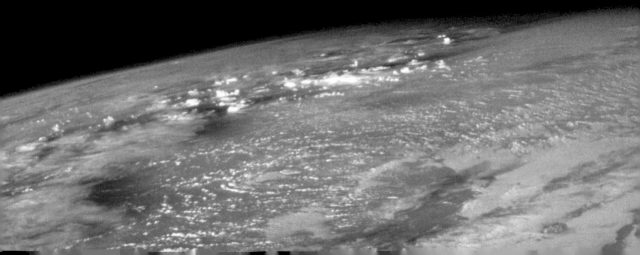

Into *Orbit*

This chapter's gallery places Hubble in orbit and showcases its improved imaging capabilities after several repair missions.

3.9.02 Deployment of Hubble after servicing mission 3B.

2.10.03 Dumbbell Nebula detail showing knots of gas and dust larger than Pluto's orbit.

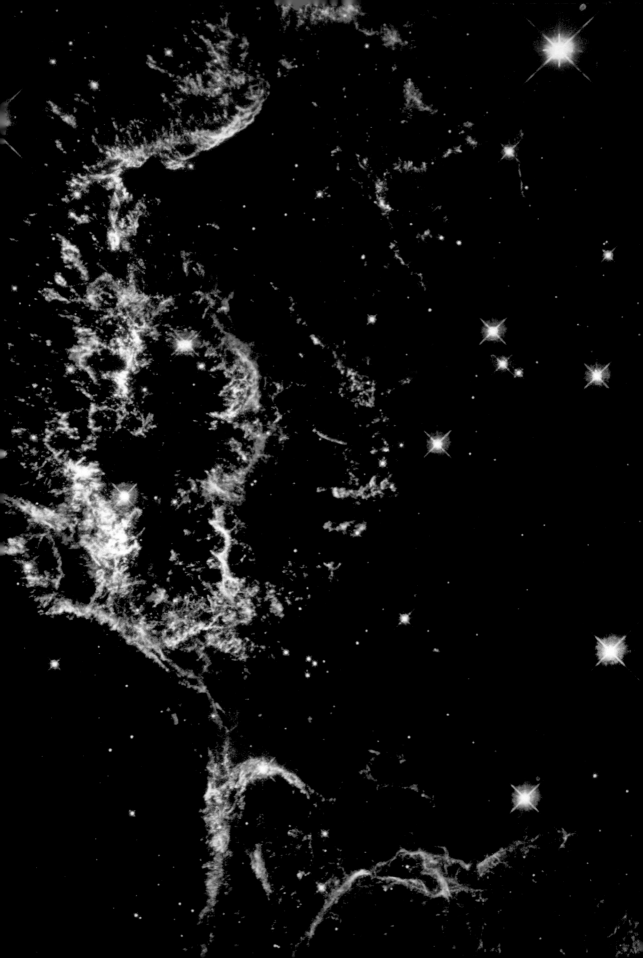

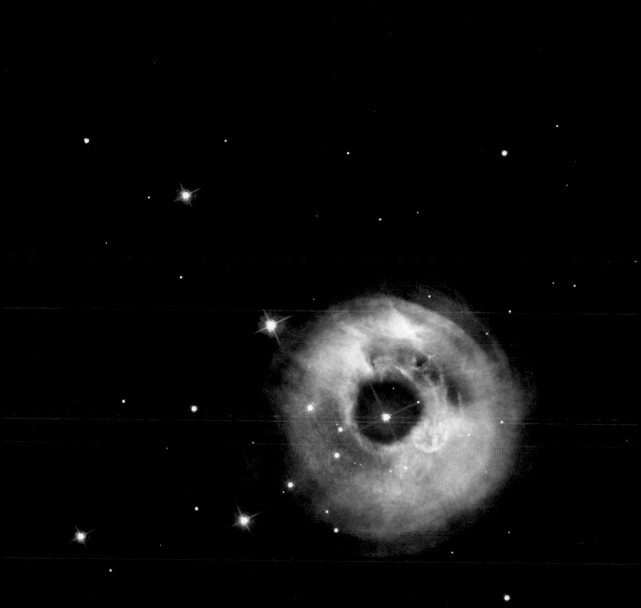

LEFT: **7.3.02** Remnant from a 1667 supernova in Cassiopeia.

ABOVE: **5.20.02** ACS image of light
echoes from an erupting supergiant star.

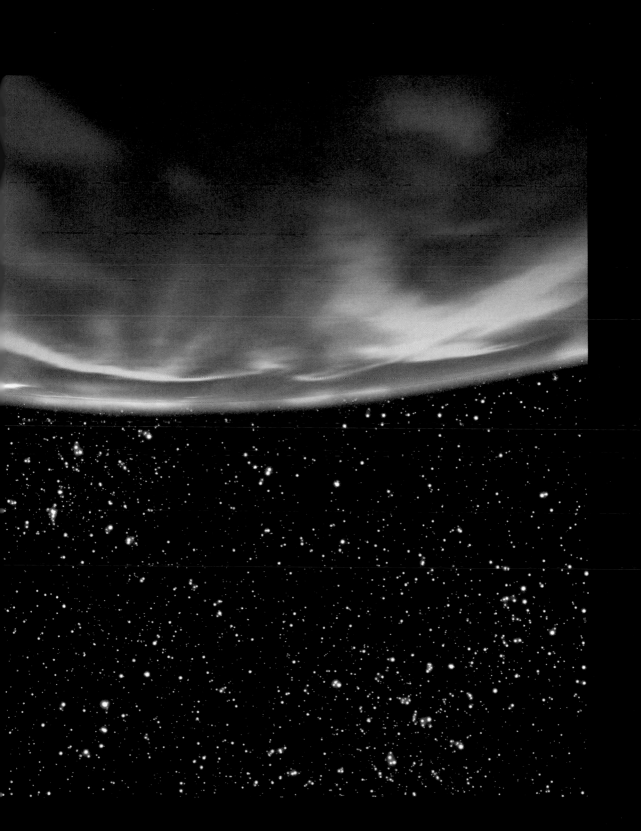

Artist's concept, from HST data, of a
planet orbiting a double star system.

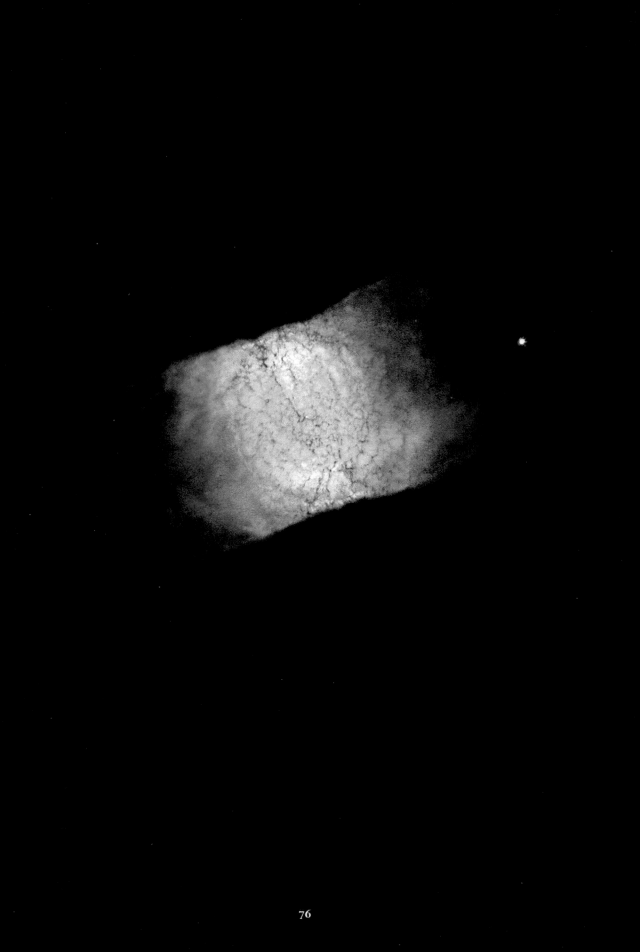

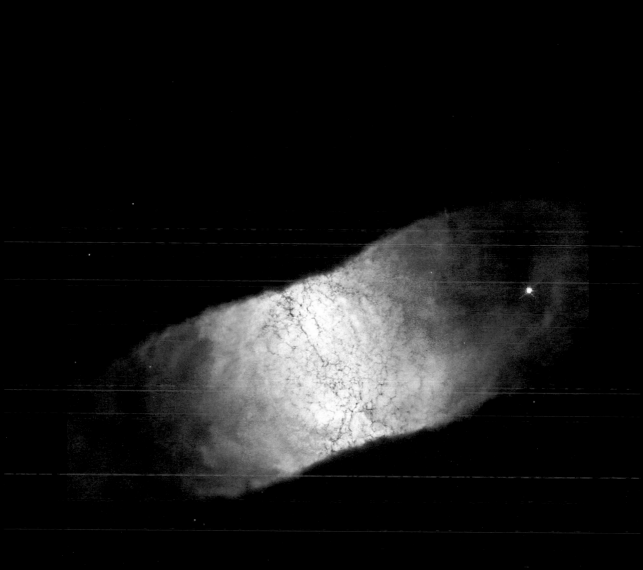

LEFT: 6.13.02 Remnant material being ejected from doughnut-shaped Retina Nebula.

ABOVE: 6.13.02 Same edge-on view of the dying star Retina Nebula in black and white.

2.97 Mark Lee and Steve Smith changing Hubble's instruments during a servicing mission.

Into Orbit

DESIGN AND CONSTRUCTION

THE CONSTRUCTION PHASE OF THE HUBBLE SPACE TELESCOPE BEGAN IN 1978. THIS VAST, COMPLEX EFFORT INVOLVED THOUSANDS OF PEOPLE AND HUNDREDS OF INSTITUTIONS, COMPANIES, AND UNIVERSITIES ACROSS NORTH AMERICA AND EUROPE. THE FUNDING BATTLES OF THE MID-1970S LED NASA TO ADOPT WHAT WAS CALLED A LOW-COST APPROACH TO BUILD HST.

Because the HST would be serviced by the Space Shuttle, the space agency was willing to relax somewhat its usual exacting testing procedures for components and the assembled spacecraft. If something went wrong once in orbit, visiting astronauts would be able to change the defective components and systems. NASA also planned to use the shuttle to retrieve the telescope and return it to Earth for extensive maintenance and refurbishment every five years or so.

In the fall of 1979, NASA established a cost-review team who concluded that the launch date of 1983 was already unlikely to be met and substantial cost increases were probable. Matters did not improve in 1980. Managers at the Marshall Space Flight Center battled strenuously to hold down costs, and in so doing pursued various strategies that to many astronomers appeared draconian. Proposals were advanced, for example, to eliminate the two spectrographs from the complement of instruments for the first flight and install them later in the telescope's life. If such plans were put into effect, the result would be an observatory far from the one envisaged by astronomers when construction had begun two years earlier.

An unwieldy management structure was a handicap, but the chief problem project managers, scientists, and engineers confronted was that they had too little money to perform all the tasks required. After much discussion, debates on revisions to schedules, and adjustments to budgets, in late 1980 NASA opted, as one senior manager put it, to "swallow hard" and "preserve the integrity of the program."[i] This was a tough decision at a time when NASA's main program, the Space Shuttle, itself needed additional resources (the first launch of a shuttle, *Columbia*, would take place in April 1981). HST's launch had been pushed back at least 18 months from the originally planned 1983, and the projected cost had jumped from a range of $540 million to $595 million to a new range of $700 million to $750 million.

Against this troubled backdrop, in 1981 Perkin-Elmer, the company manufacturing the optical telescope assembly, completed Hubble's most important component. The 2.4-meter-diameter primary mirror would collect light from astronomical objects and then direct it, via the much smaller secondary mirror, through a

hole in its center to the array of scientific instruments, where it would be collected and analyzed. If the telescope was to produce the superlative images astronomers hoped for, the primary's surface needed to be ground, then polished to great precision.

To reach this goal Perkin-Elmer ran through a three-step cycle numerous times in which it polished, cleaned, and then measured the shape of the mirror. Critical in this process was a device known as a reflective null corrector, used to measure the mirror surface. There was no ideal model against which to compare Hubble's primary mirror, so Perkin-Elmer created one using the reflective null corrector. With its aid, laser light created a grid or optical template that could be compared with the HST primary mirror. The Perkin-Elmer engineers knew when they had reached their goal in fashioning the primary mirror because the differences in the patterns produced by the actual mirror and the optical template of the reflective null corrector were zero, or null. After the polishing was finished, the primary mirror was pronounced by NASA and Perkin-Elmer as practically perfect and as one of the program's "crown jewels." Unknown in 1981 was that the reflective null corrector contained a flaw.

Do It Right

The mirror had been finished during difficult times for Hubble. A senior NASA manager later recalled those uncertain days of the late 1970s and early 1980s when NASA insisted that the project stay within its established budget. In his view, this decision "created an absolute disaster…the Project Manager had no choice, when the money went away [but] to start slowing down, re-phasing it, even stopping work at some subcontractors in some cases." The best technical talent was lost, and many critical elements were cut or short-changed, "like reliability processes and practices, which you would ordinarily want to keep in a space program." Vital tests in some cases were cut because of false economies. The result was that "whenever these penny-wise, pound-foolish judgments were made, Space Telescope kept getting deeper and deeper into trouble." [ii]

Following extensive reviews of the Hubble program in late 1982 and

1983, NASA decided constructing the telescope on the current plan was too risky and unlikely to work. A different approach was needed. The space agency revised radically the way it was building the HST. More testing and spare parts were added, the astronomers were given a greater say in decision-making, still more time was added to the schedule, and the budget increased by hundreds of millions of dollars. NASA had pursued a low-cost approach, but it had not worked.

The dividends of the new approach became apparent in the decision to build a second Wide Field/Planetary Camera. By 1983, project members had generally accepted that this camera, which would image the cosmological depths as well as the surfaces of planets, was first among the scientific instruments. Its eagerly anticipated CCD images of assorted astronomical objects were also expected to be major selling points for the project. But during the early years of building the telescope,

10.30.03 Artist's concept of region in Lynx galaxy cluster where starbirth is extremely vigorous.

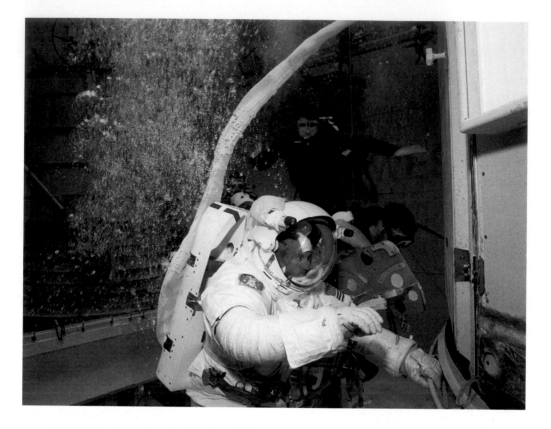

6.21.99 Mission specialist Michael Foale practicing servicing Hubble in a water tank.

various compromises had been made in its design that meant the WF/PC would be less reliable than originally planned. Because it was also the most complicated of all the scientific instruments, the calculations of WF/PC's failure rate gave managers pause. They worried that the possible loss of WF/PC and its stream of images early in HST's life in orbit would be devastating, both scientifically and politically. As insurance NASA managers authorized the construction of a second WF/PC. What became known as Wide Field/Planetary Camera 2

would possess significant enhancements over WF/PC I, most notably in its superior CCDs.

Encouraging developments also came in other areas. By 1984 substantial elements of hardware from across the United States and Europe had begun to arrive at the Lockheed Missiles and Space Company in Sunnyvale, California, for assembly into HST. Many hundreds of engineers and scientists pieced together and tested the telescope. Hundreds of others were planning for its operations in space, particularly the staff at the Space Telescope Science

Institute in Baltimore and at the Goddard Space Flight Center in Greenbelt, Maryland. They were driving for a late 1986 launch. But on January 28, at 11:39 a.m., and shortly after it left the launchpad, the Space Shuttle *Challenger* was destroyed. A joint in the *Challenger's* right solid-fuel booster failed, sending the booster crashing into the shuttle's external tank, killing all seven crew members and putting the entire shuttle program on hold.

HUBBLE TROUBLE

Shuttle flights were suspended for over three years, but at last the shuttle *Discovery* was ready for launch with HST in its payload bay in April 1990. Nearly 50 years had passed since Lyman Spitzer first envisioned such a telescope in space and 12 years after the White House and Congress had approved Hubble's construction. Many thousands of people had labored hard, and about $2 billion had been spent to reach this stage. At long last Hubble was poised to decipher the awaiting mysteries of the universe.

On April 24, *Discovery* roared into the sky above Cape Canaveral on a column of smoke and flame. All went well, and after a short time in orbit, HST was gingerly lifted out of the *Discovery's* payload bay and

placed into space, 375 miles above Earth. By April 27, the orbiter's crew had finished their tasks, and *Discovery* slowly flew away from the telescope. Hubble had become an automated observatory in the hands of its ground controllers.

The first astronomical image was of a star, returned from the telescope on May 20 by the Wide Field/Planetary Camera. There were three main features to the image: Ten per cent of the star's light resided in a small, bright core, around which was a halo as well as funny-looking features that would soon generally become known as tendrils. As an engineer put it, "This is all very encouraging for the first try.... Think what they can do with fine tuning the images. [Hubble Space Telescope] might actually work."[iii]

Yet one astronomer, Roger Lynds, had serious reservations. In his view the star's appearance was odd, and at a project meeting he voiced the opinion that it was due to one of the most common sorts of telescope aberrations, spherical aberration, which results in blurry images. Soon other astronomers, particularly Chris Burrows at the Space Telescope Science Institute, were reaching the same conclusion. At first there seemed to be a possible easy fix. Astronomers clung to the notion that if spherical aberration was present it could be readily compensated for by use of the actuators behind the primary mirror. Perkin-Elmer had designed the actuators to impart small forces to different parts of the mirror so that its shape could be altered very slightly. But at a project meeting on June 19, a representative from Hughes Danbury Optical Systems (the company that had recently purchased the mirror's maker, Perkin-Elmer) pointed out that if spherical aberration indeed existed, the actuators would be able to remove only a small portion of the problem. One astronomer at the meeting reckoned, "This is the moment we find out that we are doomed to failure."[iv]

When further tests confirmed that the HST suffered from serious spherical aberration, NASA decided to make the news public. Hapless NASA managers, engineers, and scientists faced baffled reporters at a briefing at the Goddard Space Flight Center on June 27, 1990. Instead of singing Hubble's praises, they had to concede that the telescope's primary mirror was built to the wrong shape and so suffered from a major defect—severe spherical aberration. Many members of Congress were outraged. "Hubble Trouble" appeared in the media. Spherical aberration so

alarmed some astronomers that there now seemed the nightmarish but real possibility that NASA or the Congress might cancel the project.

A time-bomb had been put into place during the polishing process. The flawed primary mirror meant that as soon as Hubble began observations with what NASA, its contractors, and astronomers had claimed was a superb optical system, it was bound to produce blurry images and provoke a political explosion. What scientists had no inkling of in 1981, nearly a decade before launch, when the work of polishing the mirror was completed, was that the reflective null corrector used to measure the shape of the primary mirror had not been assembled properly. As astronomers and engineers painfully reconstructed the process in 1990, the lens in the reflective null corrector had been inserted about 1.3 millimeters (approximately 0.05 inch) out of position. To position the lens the Perkin-Elmer engineer had used a measuring rod, over which was placed a metal cap. To be sure the rod was in the right place, a light beam had to be directed onto the end of the rod, and its reflection would check its position. A hole had been drilled in the center of the cap to reveal a portion of the end of the rod, sufficient to provide the actual

12.27.99 Astronauts working on HST during the third servicing mission.

Saving Hubble

The First Servicing Mission

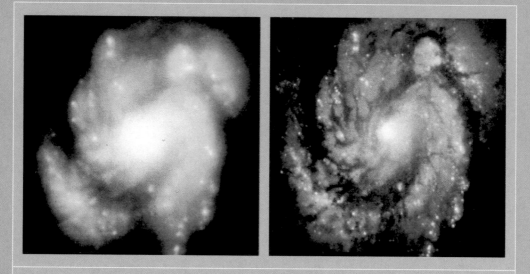

LEFT: Galaxy M100 before the first Hubble Servicing Mission.
RIGHT: 1.13.94 After servicing using the newly installed WF/PC 2.

HST was the first space observatory designed to be maintained and refurbished in orbit. Besides the modular design of its scientific instruments and Fine Guidance Sensors, other components of the telescope were Orbital Replacement Units (ORUs), built to be accessible to astronauts on spacewalks.

Following the discovery in June 1990 that Hubble was hobbled by spherical aberration, astronomers and engineers examined various schemes to fix the problem. Senior NASA managers finally settled on a shuttle mission that would push the limits of maintenance and servicing. Astronauts would fly to HST aboard a space shuttle, don their spacesuits, and install a special optical device called COSTAR (Corrective Optics Space Telescope Axial Replacement). They would also install Wide Field/Planetary Camera 2 in place of Wide Field/Planetary Camera 1. To optimize Hubble, astronomers and engineers also pressed for other changes. They wanted to remove the two existing solar arrays and replace them with a new set to fix a problem known as jitter that on occasion caused the telescope to shake by a tiny but perceptible amount. In the final plan for the mission the astronauts were assigned 11 servicing tasks.

Managers decided that a record-breaking five spacewalks would be needed with the capability for an additional two if necessary. The various on-orbit tasks were meticulously rehearsed. The astronauts trained extensively in the Neutral Buoyancy Simulator at the Marshall Space Flight Center in Huntsville, Alabama, a 1.6-million-gallon water tank that contained a full-scale mock-up of the telescope, as well as a similar tank at the Johnson Space Center near Houston. During training at the Goddard Space Flight Center in Greenbelt, Maryland, the astronauts handled the hardware they would carry into space. The crew also practiced virtual-reality computer simulations to develop a better sense of what the tasks would entail in space, marking the first time such simulations were used in training for a shuttle mission. The astronauts rehearsed operations on frictionless floors at the Johnson Space Center in order to simulate working in reduced gravity with the various components they would use. Crew members even visited the National Air and Space Museum in Washington, D.C., to work with the museum's full-size model of HST. Parts of two 1993

flights of the shuttles *Endeavour* and *Discovery* were also devoted to evaluating the designs of tools that astronauts would need to service Hubble.

When the mission began with the launch of the shuttle *Endeavour* from Cape Canaveral at 4:27 p.m. EST on December 2, 1993, it was one of the most eagerly anticipated missions in the history of human spaceflight. Once in orbit, *Endeavour* pursued the ailing telescope for several orbits. It then rendezvoused with Hubble. The astronauts captured the observatory and placed it in the cargo bay. Because Hubble was designed to be serviced in orbit, features such as handrails and foot restraints were built in to the telescope. These features would aid astronauts performing repairs in the cargo bay as the shuttle orbited the Earth at 17,500 miles per hour. Three of *Endeavour*'s astronauts assisted from inside the orbiter; the other four performed a series of spacewalks over the course of several days.

The complex mission went almost flawlessly. The spacewalks took place in full public view, made possible by sophisticated broadband television feeds from orbit. When images from the repaired HST began to be returned to Earth, all concerned could see that the repairs had worked. Hubble's vision problem had been corrected.

reflection from the rod. The cap had been painted, but a small piece of paint had become detached. The light beam directed at the end of the rod hit the exposed bit of metal on the cap instead and was reflected back. Perkin-Elmer had been measuring the position of the top of the cap and not the end of the rod.

This was a critical misstep. Even so, if there had been cross-checks in the program to catch such a mistake, it might have been corrected. There were not. With a strained budget and no expectation of additional money for extra tests, no alarm bells had been rung by the very small number of people who knew of data that indicated there might be a problem. The result was a misshaped primary mirror. It was too flat at its edges by only 2 microns, or about 1/50th the thickness of a sheet of paper, but in the world of precision optics that was a gross blunder.

Astronomers, managers, and engineers on Earth in 1990 had to live with the consequences of spherical aberration. Project members attempted to keep their focus on maximizing the scientific results from the telescope, despite its hobbled state. Even before the public announcement of spherical aberration, a number of astronomers were eagerly investigating what could be done through careful computer processing of HST's images. There were ways to subtract a significant part of the halo around each star image caused by the light that was not well focused. Astronomers applied sophisticated image-processing techniques and demonstrated in the summer and fall of 1990 that HST was still capable of producing significant discoveries. A headline in the *New York Times* at the end of August announced, "First Hubble findings bring delight," and recounted "surprising and puzzling observations" of a galaxy's center.[v] By 1992, a report in *Time* magazine on "The Best of Science in 1992" proclaimed, "The latest take: it's not perfect, but even a nearsighted Hubble is pretty powerful."[vi] But it had achieved at best partial success. As one astronomer put it: "We are getting great science from the Space Telescope despite its problems…. Unfortunately, we are not getting all of the science we paid for. We wouldn't have paid $2 billion for the capabilities it has."[vii] Clearly, the spherical aberration had to be fixed, but how?

A number of astronomers and engineers advocated bringing HST back to Earth to replace the mirror. Such a scheme, however, would cost many hundreds of millions of dollars, and some astronomers worried

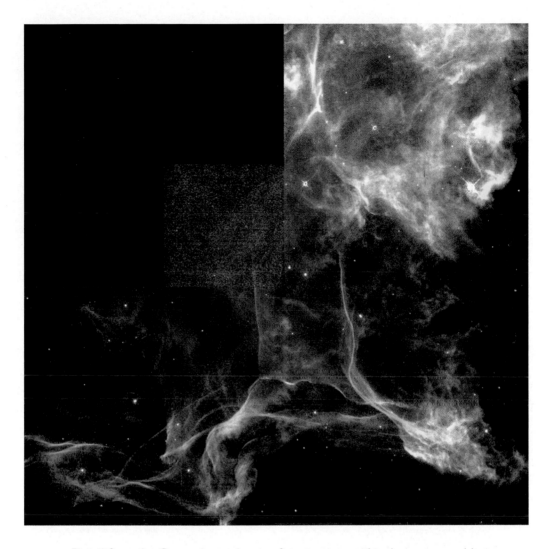

2.20.95 Detail from the Cygnus Loop showing fine structure within the supernova blast wave.

that if the HST was returned, it might never be reflown. Others preferred an alternative scheme, and ultimately it became NASA's chosen means of fixing the problem. They would build a device known as COSTAR (Corrective Optics Space Telescope Axial Replacement). COSTAR would in effect provide Hubble with spectacles. It would be shaped like a scientific instrument, and astronauts would fly it to orbit to replace one of the complement of instruments already aboard HST. It would contain a set of mechanical arms with mirrors on them that, once COSTAR was in place inside HST, would swing into exactly calculated positions so that light entering three of the remaining scientific instruments would bounce off its mirrors. Precisely curved to

12.19.99 Night launch of shuttle *Discovery* for servicing Hubble.

compensate for the primary mirror's spherical aberration, COSTAR'S mirrors would thereby restore the vision of these three instruments.

Yet NASA would not try to improve the vision of the original Wide Field/Planetary Camera. Taking another route, the space agency chose to press ahead with its replacement, the Wide Field/Planetary Camera 2. Knowing it too would display the effects of spherical aberration unless action was taken, astronomers and engineers redesigned WF/PC 2's optical components to counteract the primary's spherical aberration. Once WF/PC 2 was installed in Hubble, well-focused images would therefore fall on the camera's light detectors despite the flawed primary mirror.

MECHANICS IN SPACE

Installing the new camera in Hubble as well as making other repairs to the telescope meant planning a very complex and highly demanding space shuttle mission. The astronauts would need to capture the HST in orbit and bring it into the payload bay, then perform spacewalks over the course of several days to replace the old components with new ones. If the mission did not go well, the telescope's future would be in jeopardy. Indeed, it seemed that

NASA's own future was in doubt. News stories suggested that many lawmakers on Capitol Hill would take a dim view of NASA's ability to assemble a space station in orbit—NASA's flagship project—if the space agency proved unable to repair HST.

The stakes were therefore very high when the space shuttle *Endeavour* blasted off from Cape Canaveral in December 1993. The shuttle mission, however, went remarkably smoothly, with a number of major changes to the telescope put into effect during the course of spacewalks by the *Endeavour*'s astronauts. WF/PC 1 was removed and WF/PC 2 installed in its place. COSTAR was substituted for the scientific instrument that had been the least used by astronomers, the High Speed Photometer. New solar arrays were installed too with the aim of correcting a jitter problem. This malfunction was noticeable whenever the HST moved from very warm sunlight into frigid night when Earth eclipsed the sun. The changes in temperature caused the original solar panels to wobble a little and shake the spacecraft by tiny but significant amounts.

The *Endeavour* crew had performed their roles flawlessly, but as the orbiter returned to Earth the crucial question for astronomers was "Would the fix work?" An excited and concerned crowd of astronomers anxiously awaited the first astronomical images from the upgraded telescope, gathering at the Space Telescope Science Institute to witness the first star image emerge on a computer screen. A cheer erupted when it became clear that most of the light was concentrated into the star image's core. The spherical aberration had effectively been eliminated.

As before-and-after images that marked the major improvement in HST's performance appeared in newspapers and magazines and on television, public perception of the telescope rapidly began to shift. No longer was Hubble synonymous with trouble. Another key step in the public rehabilitation of HST came in 1994, when it was used to observe the approach to Jupiter of a fragmented comet, in addition to the effects of these fragments after they smashed into the planet. The collisions drew tremendous interest, and the HST's extremely public role in observing them underlined its capabilities. A spectacular WF/PC 2 image of a recent supernova remnant, taken in early 1994, which appeared to show a system of rings and haloes, also won wide publicity for the Hubble.

These and other images positively engaged the general public and garnered applause from the telescope's patrons in NASA and ESA, on Capitol Hill, and in the White House. HST was at last performing up to the standards originally envisaged for it. Provided no insuperable technical difficulties arose, the telescope's life was now guaranteed, at least for the near future.

By the time of the first servicing mission to HST in late 1993, two new scientific instruments were well on the way to completion, and both were transported to HST during the second space shuttle servicing mission in February 1997. The Space Telescope Imaging Spectrograph (STIS for short) was designed to replace many of the capabilities of the two original spectrographs and was a more powerful instrument in many respects than both combined. The Near Infrared Camera and Multi-Object Spectrometer, or NICMOS for short, was designed to observe objects in the near infrared part of the electromagnetic spectrum—that is, light of a wavelength invisible to the human eye—and provided a powerful complement to WF/PC 2. In order to detect infrared light from astronomical targets, not from the heat radiation created by its own electronic systems, NICMOS's detectors needed to operate at extremely cold temperatures. For this reason they were fitted inside a thermally insulated cooler somewhat like a thermos bottle, which contained a 230-pound block of nitrogen ice, there to serve as the coolant. Unfortunately, a problem soon developed after NICMOS was installed in the telescope. Instead of a lifetime of four and a half years, NICMOS was depleted of coolant after only two, running out by January 1999. The instrument was effectively out of commission. When the fourth shuttle servicing mission was launched to the telescope in March 2002, the astronauts of the space shuttle *Columbia* carried to orbit a specially designed NICMOS Cooling System. The astronauts successfully inserted the device, which operates much like a household refrigerator, and it chilled NICMOS to below a very frigid minus 321 degrees Fahrenheit.

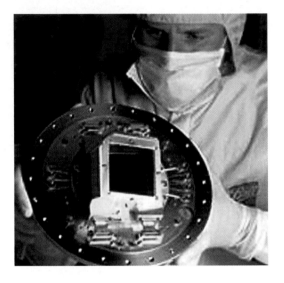

16 megapixel CCD for WFC 3, capable of improving sensitivity by a factor of 10.

NICMOS was brought back from the dead and happily performed even better than it had in its first incarnation.

During the same 2002 shuttle mission, astronauts installed in one of the scientific instrument bays behind Hubble's primary mirror a new and highly efficient camera, which was called the Advanced Camera for Surveys (ACS), replacing a now obsolete instrument. With a field of view twice as large as WF/PC 2, and able to observe in the ultraviolet, visible, and near infrared regions of the spectrum with larger, more sensitive detectors giving twice the image sharpness, ACS added much more observing capability to the Hubble Space Telescope.

By March 2002 HST was in many ways a different observatory from the one first launched in 1990. All of the original scientific instruments had been replaced by more powerful ones. The telescope's support systems on Earth and in orbit

12.99 Earth and Shuttle in an astronaut's visor during third servicing mission.

had received major upgrades over the years, too. In the end, Hubble's scientific performance had vastly exceeded the hopes that had sustained its advocates over so many years. By early in the 21st century, after over a decade in space, it had come to be widely regarded as the most productive observatory ever built. With even more instruments planned for later shuttle missions, users of Hubble Space Telescope could look forward to an even more capable telescope able to add to its reputation as a scientific and engineering triumph.

On January 16, 2004, in the wake of the Bush Administration's call for a shift in NASA priorities, NASA Administrator Sean O'Keefe announced that the planned servicing mission 4 to Hubble would be cancelled. His decision came under Congressional review as Hubble continued to provide unprecedented new views of more and more distant galaxies, and in March released the first results of the ACS Ultra-Deep Survey, exploring a time when galaxies were just beginning to form. In limbo are two new instruments, including a new Wide Field Camera 3 and a new spectrograph, which promise to enhance Hubble's performance by yet another order of magnitude.

Observing
with Hubble

IN THIS CHAPTER:

This chapter's gallery showcases studies of spiral galaxies made by the Hubble Space Telescope.

11.4.99 Tidally distorted interacting galaxies in Canis Major.

LEFT: 5.25.99 Spiral galaxy in the
Centaurus cluster of galaxies.

ABOVE: 9.14.97 An unusual coincidence of
two unrelated galaxies in Hydra.

6.3.99 A flocculent spiral galaxy in Coma Berenices, part of the Hubble Key Project.

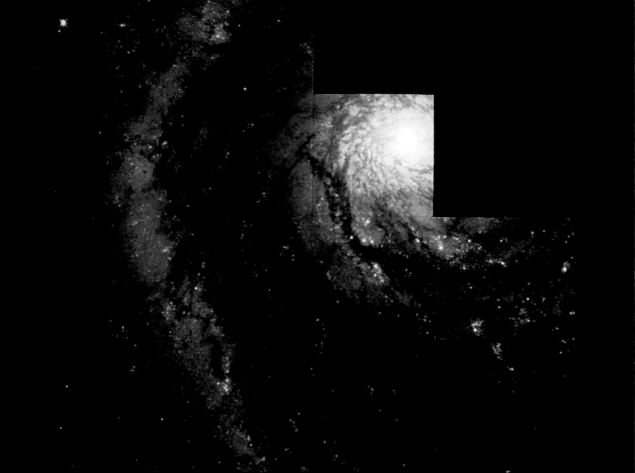

LEFT: 5.9.96 Central region of M51 with the WF/PC 2 mosaic from all CCD fields.

ABOVE: 5.9.96 Ground-based image of M51 with WF/PC 2 field superimposed.

3.9.02 Hubble after release from *Columbia* on the 3B servicing mission.

Mosaic CCD image of M51 from Kitt Peak National Observatory's 0.9-meter telescope.

Observing
with Hubble

The Structure and Evolution of Galaxies

SINCE THE 1980S NICHOLAS ZABRISKIE SCOVILLE HAD BEEN INTERESTED IN HOW STARS FORM IN THE SPIRAL ARMS OF GALAXIES. NOW A CALTECH ASTRONOMER, HE HAD USED RADIO TELESCOPES IN MASSACHUSETTS AND THE OWENS VALLEY IN CALIFORNIA, STUDYING THE BEHAVIOR OF GIANT MOLECULAR CLOUDS IN BOTH THE RADIO AND INFRARED REGIONS OF THE SPECTRUM.

By the mid-1990s, he decided to study the distribution of hot hydrogen gas surrounding very hot stars in the central portions of spiral galaxies, and the Whirlpool Galaxy became his primary target. He assembled a team of similar-minded colleagues, including Maria Poletta from the University of Geneva; Sean Ewald and Susan R. Stolovy, staff members at Caltech; and Rodger Thompson and Marcia Rieke, two astronomers from the Steward Observatory in Tucson, Arizona. Thompson was principal investiga-

tor for NICMOS, which would be placed on Hubble in the second servicing mission in February 1997, and Rieke was also very active in NICMOS. In the spring of 1996 the team applied to the Space Telescope Science Institute for telescope time as a General Observer, or GO, to explore the spiral structure and formation of the hottest stars in the Whirlpool, known as M51. They wanted to produce high-resolution images in both broad and narrow ranges of the visual and infrared spectrum to determine the detailed

10.17.94 Generations of stars in superimposed clusters in the Large Magellanic Cloud.

structure of dust lanes within the spiral arms. They would combine their observations with radio maps of the same regions "to determine local physical conditions."[i] Members of his team also proposed to observe M51 with NICMOS, and won two observing runs as Guaranteed Time Observer, or GTO, because they were members of the instrument team.

The astronomers were interested in the larger problem of what trig-

gers the formation of stars in active galaxies such as the Whirlpool. Scoville regarded the Whirlpool as a Rosetta Stone for such work: It is a comparatively nearby galaxy, its face-on orientation reveals its grand-design spiral pattern, and the amount of gas and dust lurking within its magnificent arms provides ample material to make new stars. The gas and dust itself is a mix of the original stuff that formed the galaxy, plus younger materials formed in the hearts of massive stars that have already lived and died in the galaxy's history. "From these ashes," Scoville later stated, "the future generations of stars will arise."[ii]

Ultimately, from this observing run, a contribution would be made to the Hubble Data Archive, which serves as the repository for collected data. In this manner, the data gathered will be of continued use to astronomy long after Scoville's team has completed their analysis.

Nick Scoville's proposal was certainly very intriguing, and he and his team had the combined experience and know-how to make an effective run. But this is the case with the vast majority of observation proposals. How was it then that the Hubble Space Telescope was pointed toward M51 to catch those red photons on July 21, 1999 at 4:30 in the morning (EST)? Observation time is open to anyone who is capable of submitting a qualifying proposal. The key in getting time is persuading the granters that the goals of the proposal are worth the sacrifice of other potential goals. Observing time is restrictive, and for every individual or team that succeeds in getting a run, many others are left without the use of Hubble's very significant "eyes." Scoville's team would win this bat-

tle, but would have to wait two years for the privilege of a run.

THE PHASE I PROPOSAL

The Space Telescope Science Institute asks astronomers on a regular basis to submit proposals to use the telescope. Each time such a call is issued it starts a new annual observing cycle. There are two phases to the proposal process. In Scoville's case, he knew that his team's Phase I proposal was due at the Institute by September of 1996. They needed to fashion as strong a case as they could for the scientific importance of the observations they wanted to take in order to convince peer reviewers that the request justified the use of the HST. Scoville's team also had to predict reasonably well how much observing time, measured by the required number of telescope orbits around the Earth, would be needed by one of HST's instruments to meet their goals.

The team's first hurdle was to gain the approval and support of a subject-related Proposal Review Panel (PRP), eight astronomers who convene at the Space Telescope Science Institute to examine proposals in a particular specialty such as galaxies and clusters of galaxies; hot stars; cool stars; the interstellar medium; or solar system objects. The chairs of each PRP then carry their reports to a second-tier review group called the Time Allocation Committee (TAC). The TAC for Cycle 7 met on three days in late November. At the end of December, Scoville and his team found out that they had secured 280 minutes of precious HST observing time with WF/PC 2, equivalent to the time taken by Hubble to complete just under three orbits, among the larger amounts of observing time granted to GO category proposals.

THE PHASE II PROPOSAL

The reward for those who survived Phase I was to provide a far more detailed proposal to the STScI two months later. Now the team, with the assistance of Institute staff, had to identify specific requirements for virtually every second of the observing run. Working at this level of detail would help the staff responsible for HST operations to integrate the proposal into the observing schedule as efficiently as possible. The Institute staff would translate Scoville's proposal into a

Kitt Peak National Observatory provides state-of-the-art telescopes to all astronomers.

sort of libretto that would identify every step in the process: locating the guide stars that would direct Hubble to the different portions of M51 he hoped to record, calculating the required exposure times, determining the order of the filters they would observe with, and planning the most efficient ways to process the data. This position resembles that of an electronic orchestra conductor who has to plan in advance for every nuance in a piece of music.

Each exposure with HST has to be meticulously planned out as there are definite limits to where the telescope can be pointed at any particular time. The staff of the Institute must account for these constraints when they construct the observing schedule. This is usually done by taking the coordinates of the objects the observer wants to study and entering them into a huge computer processing program that includes the orbital parameters of the HST, the Tracking and Data

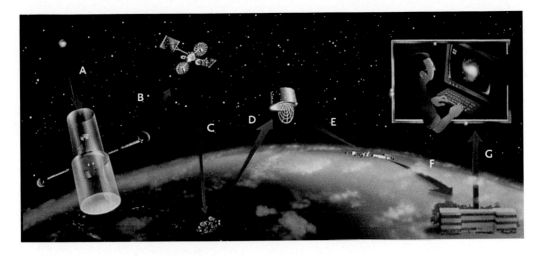

Star light enters telescope (A). Hubble sends signal to TDRSS (B),
which is then relayed to ground station (C) and beamed to low Earth orbit satellite (D)
then to NASA/Goddard (E) and on to STScI (F) where it is stored and processed (G).

Relay Satellite System (TDRSS), the Earth, the sun, and the moon. The program then develops a detailed schedule that helps the telescope staff plan out every step of the observing run. This preparation lies at the heart of a Phase II Proposal.

The observatory must work within certain physical parameters. HST is in low Earth orbit, only 375 miles (600 kilometers) above its surface, not even in true space. At its current altitude, HST completes an orbit every 97 minutes at a speed of 17,500 miles (28,000 kilometers) per hour. Given its varying position, it must never be pointed toward the Earth, the sun, or the moon. The sun is so bright that in normal operations HST rarely looks closer than 50 degrees in its direction. If it drifts to within 20 degrees of the sun, the Hubble is designed to close its aperture door so that no light can enter the telescope tube. Plus, there are also parts of the orbit where intense magnetic fields demand that the telescope partially shut down to avoid damaging the sensitive detectors and other electronic components.

While the optics of the telescope must always avoid the sun, its housekeeping elements, such as the solar panels that generate the power the telescope requires, need the sun. Also, two radio antennae must always be able to connect with the Tracking and Data Relay Satellite System (TDRSS) in higher Earth orbit, or with one or more ground stations. All of these considerations must be taken into

The Whirlpool Galaxy

From the 1800s

The spiral pattern the Earl of Rosse drew in the 1840s that so fascinated everyone led to its being named the Whirlpool. To make these drawings, however, was not an easy process. An observer could not just press his eye to the telescope's eyepiece for a few minutes and then record what he saw. Rather, one such drawing was composed over a series of many nights in which different details were discerned and added to the evolving rendition. On any one night the object could be examined for only an hour or two, when it was near enough to the meridian of the sky (due south) to be accessible to Rosse's highly restricted telescope. Additional steps were required to transfer the image to journals and books, for these steps required using engravers, each of whom had a particular style and left his mark on the work.

There are striking parallels between Rosse's efforts to capture the shifting views of M51 with pencil, ink, and paper, and Nick Scoville's team as they sought to bring out the key features of the Whirlpool Galaxy. The images from Hubble are definitely not "snapshots" or single pictures. An extensive processing sequence is needed to make electronic images useful for science, as well as appealing to all: flattening or adjusting the scale of the images to create mosaics; adding and modifying color to highlight physical properties; cleaning up cosmic rays to reduce confusion; stretching light values in the middle ranges of the exposures to see more detail; orienting and cropping to direct the eye to important details. All these steps are done with varying degrees of judgment, intuition and precision, just as Rosse labored to impart his vision through the engraver's hand.

As both the Herschels and the Rosses well knew, a nebula can present a very different appearance when studied with the same telescope on different nights. The behavior of the Earth's atmosphere was not consistent from night to night (or even

The third Earl of Rosse's six-foot speculum mirror reflector, affectionately nicknamed the Leviathan.

minute to minute). Big reflectors were notoriously idiosyncratic and did not act perfectly consistently. Maybe the mirrors were a little more tarnished or the primary mirror had been repolished. The observer's vision too might vary somewhat from night to night. Moreover, for wide dissemination Rosse enlisted artists and engravers to translate his drawings into a form suitable for publication. He keenly knew that this step in the process changed the nature of his observations. Thus, published drawings and engravings of a nebula, by their very nature, could not provide a completely faithful representation of what an observer had witnessed at the telescope. In effect, the drawings and engravings were composite images, the synthesis of a range of observations and considered judgments, certainly not a simple snapshot of Charles Messier's M51 at one time. A very similar series of decisions were made by Nick Scoville and his team. Much has changed in astronomy as its technology has changed, yet so much has stayed the same.

account when planning the details of the proposal.

THE OBSERVING RUN

Astronomers today spend very little time actually looking through telescopes. No longer do they work in darkness with a huge, slow-moving apparatus. Computers, video technology and remote control systems automate observing routines using electronic photography, spectroscopy, and photometry, turning telescopes into robots. Instead of braving the frigid night air, astronomers now typically observe in secure, lighted and heated office settings at the observatory, or even at home. And it makes little difference where the telescope is located: in the next building, around the world, in low Earth orbit (like HST) or even 23,000 miles overhead in Earth-synchronous orbit. Observing today is also a far more efficient process. Short visits to the telescope, either physically or virtually, called observing runs, have replaced unending observing seasons.

An observing run on HST is usually a fully automatic, programmed process. For this reason, once the observing schedule is established, almost endless little details have to be thrashed out ahead of time in Phase II planning. Anywhere up to a dozen observing projects may be contained in one set of commands, so the program combines them into a seamless continuum with the aim of wasting as little time as possible in shifting the telescope to new parts of the sky. The less time wasted, the more time that can be spent observing astronomical targets.

One important consideration is choosing the appropriate guide stars. These must be identified well ahead of time and their positions carefully programmed into the observing schedule. They should be bright enough for the Fine Guidance System to lock onto, but not so bright that they do not allow for sensitive adjustments to be made in the telescope's position. These guide stars have to fit within very well defined arc-shaped regions called pickles in the field of view of the Hubble. Guide stars are found using a giant digital sky survey maintained at the Institute, based upon the classic Palomar Sky Survey. The HST pointing system requires star positions to be accurate to 0.3 arc second, or over 6,000 times smaller than the disk of the full moon.

The observing schedule is complete when the guide star fields are selected and exposure times calculated. It is then entered into the Institute's Science Planning and Scheduling

12.4.03 Huge star-forming region in the spiral galaxy M33 in Triangulum.

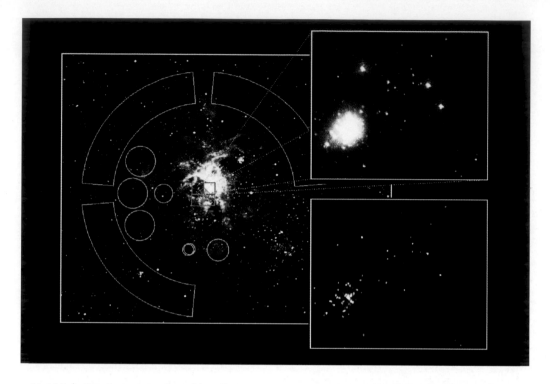

Hubble's focal plane is shared by all its instruments, including WF/PC 2 [the square] and the guide star fields [the arching "pickles"].

System, and eventually the actual sequence of commands is stored in the HST's onboard computer.

THE WONDERS OF M51

As Hubble completed a series of calibration tasks for the CCD chips in STIS—the Space Telescope Imaging Spectrograph—it was also preparing for its next program, slowly moving toward the location of the Whirlpool Galaxy, an area within the arc defined by the handle of the Big Dipper known as Canes Venatici (the Hunting Dogs). Sensitive onboard gyroscopes told the telescope when it was pointed in the right direction, causing counter-rotating reaction wheels to cease the telescope's motion. Its last observing task had been to study a galaxy in Lyra, 60 degrees east of the Whirlpool, so the slew time was taken up with the calibration tasks. When the gyros reported that the telescope had now reached the correct position, small telescopes called star trackers refined the telescope's position to less than a minute of arc (1/30th the full moon disk). At this point Hubble was ready to acquire the object field and the guide stars. Once the Fine Guidance Sensors locked on to the right guide stars, various tiny adjustments were made and every part of the system reported back all was well. The telescope went into its fine-lock mode and began to expose the Whirlpool in the first hours of July 21, 1999.

During an exposure, astronomers and technicians on duty at the Institute monitor a large bank of computer screens that display real-time data on the performance of the scientific instruments and the environment inside the spacecraft. If streams of raw data downloaded from HST show anything at all amiss, adjustments can be made in real time if the conditions are right and the operator acts quickly and decisively. One of the

4.5.01 Image of the central portion of M51 reveals spiral structure to its very core.

most serious concerns is that the telescope may loosen its lock on its guide stars and so on the object being observed—something that potentially can happen anytime because the gravitational field of the Earth is uneven, its magnetic field creates differential forces, and the tiny amount of residual atmosphere at the height of the telescope is still sufficient to

cause perturbations. Problems arise if the telescope drifts by more than seven ten-thousandths of a second of arc.[iii] The flight computer will usually instruct the reaction wheels to compensate, but the process is one that has to be watched closely.

Hubble had been instructed to take 11 exposures of M51 between 12:58 a.m. and 5:39 a.m. EST, on July 21, observing in six wavelength ranges between the ultraviolet and the near infrared with exposures lasting from 300 seconds up to 1,300 seconds. Whirlpool data were first relayed through one of the elements of TDRSS in stationary orbit to a ground station at White Sands, then to Goddard, and finally to the Institute for processing.

After traveling to Earth for 30 million years, photons from the Whirlpool Galaxy ended their lives on the four CCDs in WF/PC 2. Each CCD reported a data stream including intensities and coordinates that were stored separately

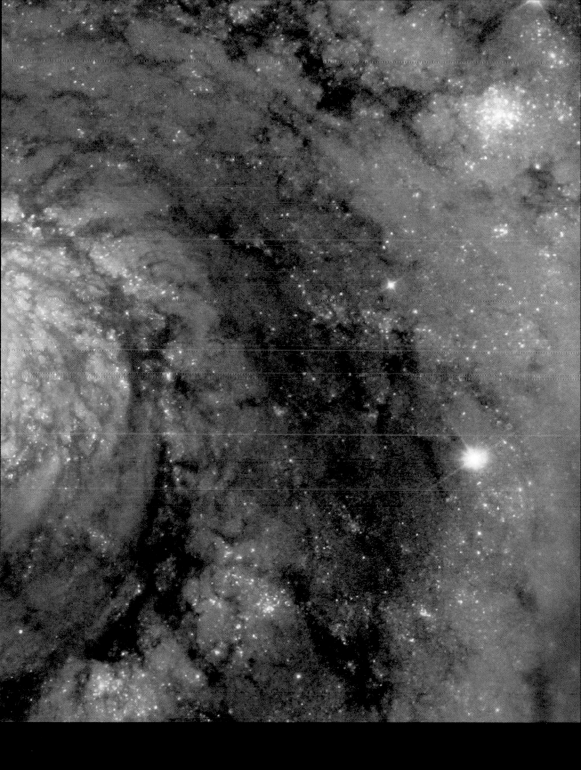

4.5.01 The same data as page 117 processed differently reveals M51's outer spiral detail.

but would later be combined into a wide-field image of a large portion of the galaxy.

INTERPRETING THE DATA

There are two major steps in bringing the data to the end user: downloading it from the satellite, and calibrating and correcting it for both systematic errors due to the instrument and random errors due to momentary peculiarities of the space environment. Data archived at the Space Telescope Science Institute are electronically rendered into a standard data format called the Flexible Image Transport System, or FITS, which every astronomer is supposedly able to read and manipulate no matter which scientific instrument produces the data.

Due to the nature of their science, astronomers have been working collectively for many decades. In the 1950s and 1960s, astronomers had to organize themselves to best use a small number of huge telescopes at national facilities, rather than a large number of small telescopes. They had to find ways to ensure that each telescope operated smoothly and continually, so that everyone could learn quickly and enjoy efficient access. Thus American astronomers created FITS in the 1970s as a means of sending

data seamlessly to one another. Unlike better-known formats, such as GIF and JPEG, FITS is optimized to deal with scientific data that are stored in complex checkerboard arrays and for tables of data where row and column information has to be preserved for efficient access. FITS was accepted by the world body of astronomers, the International Astronomical Union, at a meeting in Baltimore in 1998.

FITS, however, is only the first step in making data transportable. It must also be processed. An astronomer using FITS can calibrate and edit, adjust for systematic errors such as the flatness of an image or wavelength calibration, and even subtract background noise. But more sophisticated and comprehensive management of the data requires another level of processing. To meet the challenge, once again astronomers created a virtual software entity called The Image Reduction and Data Analysis Facility, or IRAF. The United States, UK, Australia, the European Southern Observatory, and Japan all adhere to the system, managed by the IRAF programming group housed at the National Optical Astronomy Observatory home base in Tucson, Arizona. IRAF provides astronomers with a tool to process

The Complicated Privilege

On Proposing to Use the Hubble

To gain coveted HST time, one may propose as a General Observer, or GO, and compete in a complex process. The Hubble Space Telescope is both a national observatory and an international facility. This means that anyone who submits a proposal will be considered. The proposer does not have to belong to an institution, or even have a Ph.D. in astronomy, but both are important in helping to prepare a proposal that will convince others that the goals are worth the time and effort.

In its first decade, the Space Telescope Science Institute received 8,000 proposals for time. Only a fraction were awarded. Proposals can be made by individuals, but more and more groups propose today, as problems have become larger and more complex, requiring many talents, and observing time is very limited.

Proposals fall in several classes. GOs compose the bulk and constitute the usual route to gaining time. But there are also Guaranteed Time Observers (GTOs), people who had a role in building the telescope or one of its instruments, or people who are performing critical administrative functions. Finally, there is Director's Discretionary Time, amounting to ten percent. Discretionary time can be used to accommodate fast-breaking events such as a comet or a supernova, or it can be applied to larger or longer-term programs that are desirable but not competitive in the GO category.

Observing time is broken down into cycles. Astronomers propose for a cycle at least a year ahead but typically much more, as the schedule sometimes slips. For instance, proposals for observing during Cycle 5 (mid 1995) were announced in June 1994. Two years is not an unusual waiting period.

The Institute has created a Hubble Data Archive available to anyone. In this way, Hubble data will continue to be used long after active observations cease. As with any astronomical data, recording the properties of the universe at any one particular moment, the record preserved becomes more valuable with age.

False-color image of M51 undergoing processing.

images and produce graphics as well as a standardized way to analyze all forms of electronic data in the FITS system. Major astronomical instruments, both ground-based and space-based like HST, also have specialized front-end software packages to ease the data-processing procedures for the end user. Together these constitute a data-processing package for the astronomical user that is continually updated and refined.

Armed with high-powered image-processing workstations, Nick Scoville and his team all have the know-how to manipulate FITS data using IRAF, which embodies many of the features found in commercial products such as Photoshop. They could cut and paste, crop, flatten, shift, dither, drizzle, or add color to individual images to better reveal how the particular elements emitting light in that wavelength range are distributed. These techniques can enhance contrast, not only to make various structures visible, but also to do so in a way that retains full control of the information contained in the image.

Each physical manipulation of an image, however, carries with it some level of interpretation, and requires that the astronomer, through training, talent, and intuition, retain intellectual control of the entire process. Every form of image manipulation has to be justified and rationalized among peers who ultimately arrive at some consensus about what is proper practice and what is out of bounds. Deciding where to draw the line when trying to find the continuum in a noisy spectrum or trying to establish the nature of a light curve of a rapidly varying star takes judgment. These skills with digital images are as central today to astronomical practice as knowing how to properly calibrate and develop a photographic plate a few generations ago.

Institute image processing facilities in the mid-1990s when Scoville's team gathered data.

Scoville's team first made the separate exposures covering the M51 field fit as neatly as possible to reduce distortion for the mosaic of images they were producing. They subtracted out noise and the spurious cosmic-ray tracks that appear on every image. Then with a clean mosaic, they fit the WF/PC 2 data to infrared observations they later secured from NICMOS. They added to these archived WF/PC 2 data from an earlier observing run by another group of astronomers led by Holland Ford of the Johns Hopkins University. In fact, Scoville's WF/PC 2 images were a supplement to complete what Ford had started in 1995. (By agreement, the data obtained for Principal Investigators remain proprietary for only a year. After that anyone who registers for an account to access astronomical data at the Institute can use them, assuming they have the right computer equipment. All the data from a national facility become freely

7.26.01 Radiation from a cluster of hot stars (lower center) trigger star formation in this portion of the 30 Doradus Nebula.

available after a set time.) Scoville's goal was to combine all the images from both instruments to determine the overall spiral structure of the Whirlpool arms. This structure was defined by the different components of the arms: cold dust clouds and hot, excited hydrogen shells surrounding very hot, young blue stars. These stars more often than not exist in small clusters known as associations, much like the group of brilliant stars lighting up the Orion Nebula in the Milky Way galaxy. Scoville's group had set out to perform a general reconnaissance of many Orion-type associations in the Whirlpool, doing whatever they could to link individual clusters of hot stars to their original parent dust clouds. NICMOS infrared images cut through the dust and gas to see the stars still buried within their dense embryonic clouds, providing the exact positions of where the stars are being formed. The WF/PC 2 images isolated the excited hydrogen regions and the structure of the gaseous components of the clouds around stars farther removed from their parent clouds to create an overall structure for the stars, cold dust, and hot gas.

Exploiting the high spatial resolution of the WF/PC 2 images, the team found a new structural element to spiral galaxies. Through being able to resolve structures only 12 to 30 light-years in size in the Whirlpool's arms, rather than the resolution limit of 320 light-years available from ground-based telescopes in the past, they found that the famous spiral structure of the Whirlpool was far more complex than previously imagined. Hundreds of little spurs of dust could be seen branching out from the main arms in patterns suggesting rather twisted Christmas trees. The structure of this dust component helped them rethink how the cold dust was distributed in galaxies like the Whirlpool. Most important, from all the combined data of WF/PC 2 and NICMOS they were able to pro-

duce a detailed model that describes how regions of star birth migrate through not only the many compact associations they studied, but through the galaxy itself. In so doing, Scoville and his team hope to have contributed to a fuller understanding of the evolution of a galaxy's structure over cosmic time.

The team submitted its report "High-Mass, OB Star Formation in M51: Hubble Space Telescope H[alpha] and Pa[schen alpha] Imaging" to *The Astronomical Journal.*[iv] They also prepared other scientific papers emphasizing different aspects of their ongoing research. Well before publication of their papers, however, on April 5, 2001, the world learned the gist of their effort in a press release from the Institute. The central feature of the release was the best full-color image yet of the Whirlpool, in all its magnificence, using all the data available from Hubble and adding to it images from a medium-size ground-based telescope at Kitt Peak National Observatory near Tucson, Arizona. "Hubble and Kitt Peak Take Combined Look into the Heart of a Galaxy" was posted and ran on *Space.com* by 9:00 a.m. that day. It was picked up by newspapers and weekly magazines and soon gained a life of its own.

The Wonder of Outer Space

This chapter's gallery showcases the range of nebular forms imaged by HST.

4.24.03 Detail of the Omega Nebula in Sagittarius, where stars are being born.

LEFT: 7.3.03 Debris from a supernova surrounds an exotic form of neutron star.

ABOVE: A dying star ejecting material at a fierce rate.

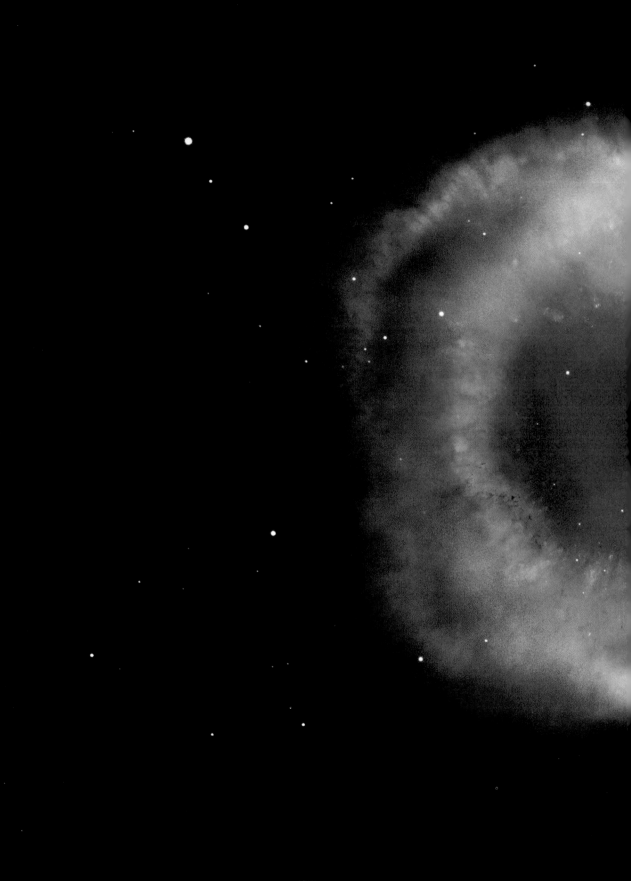

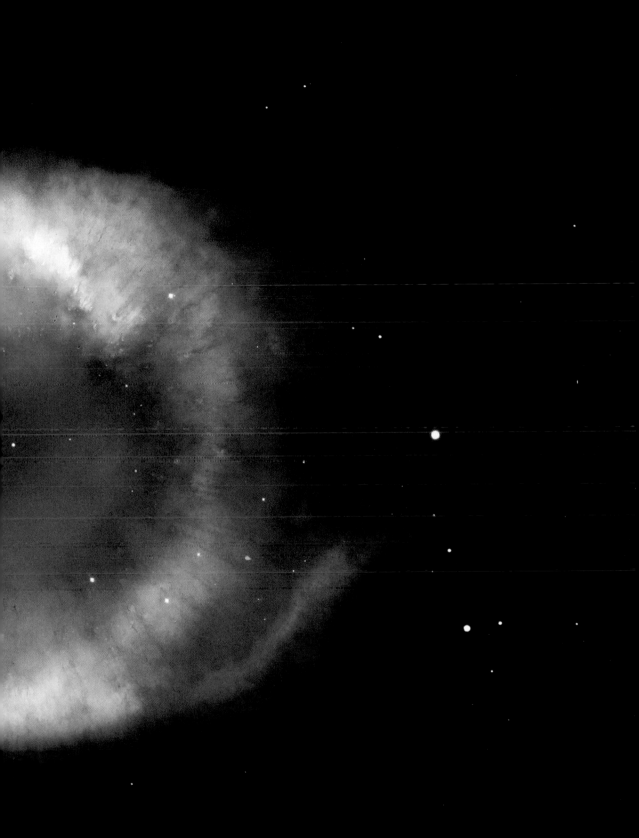

5.9.03 Mosaic image of the famous Helix
nebula, released on Astronomy Day.

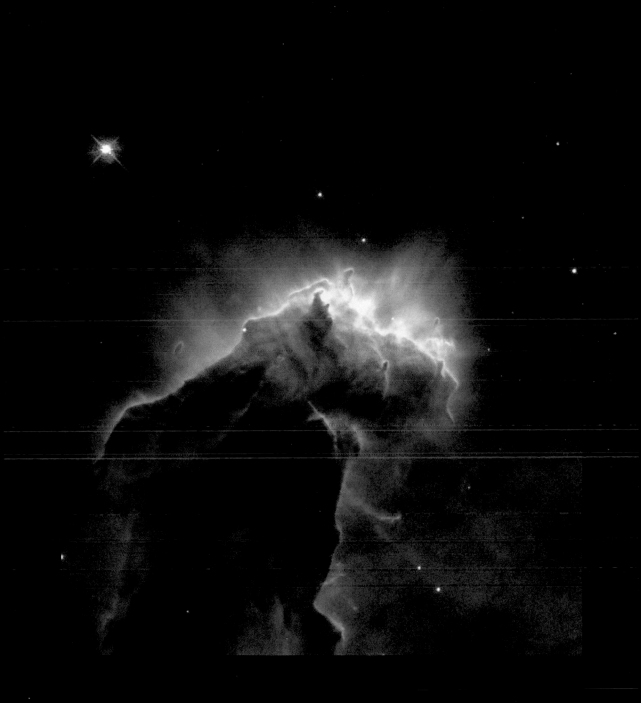

LEFT: 11.2.95 Monochromatic image of a portion of the Eagle Nebula.

ABOVE: 11.2.95 Superposition of multiple monochromatic images through filters yields a color image.

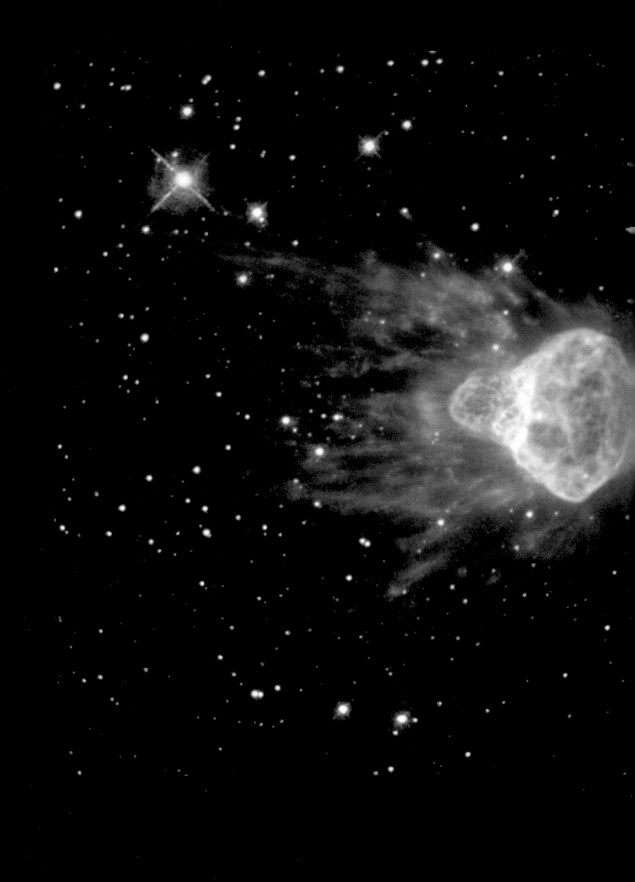

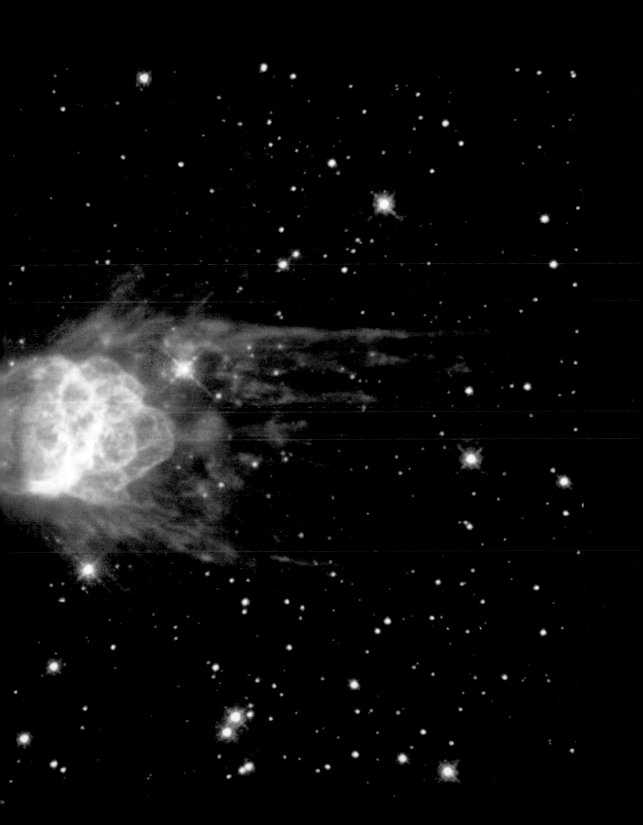

2.1.01 The Ant Nebula's shape is possibly due to intense magnetic fields.

4.30.02 The Cone Nebula, an intense region of star formation in Monocerus

The Wonder
of Outer Space

To create the view of the Whirlpool Galaxy as it appeared in *The Astronomical Journal*, astronomers assembled several views in a mosaic to expand the camera's field of view, combined exposures to create color, and eliminated some of the most obvious defects of the detector. Further processing was done before the image was released as part of the Hubble Heritage Project. Image processors combined the WF/PC 2 data with ground-based observations, subtly adjusted the colors, changed the contrast between the galaxy's core and its arms, and shifted the orientation. The resulting image shows a well-known astronomical sight in an even more impressive manner.[i]

The Hubble Space Telescope was designed principally as an imaging instrument. But only after astronomers, administrators, and public outreach officials saw the images it was finding did they develop an appreciation for their aesthetic power. Remarkable as it may seem today, no early release images were planned when the telescope was first launched. Both NASA managers, who were concerned about the response if HST should malfunction, and some astronomers, who wanted to ensure they would be the first to use this remarkable new instrument for science on particular targets, argued against immediately showing pictures to the public. Over the lifetime of the telescope these positions would change, and most would come to recognize the popular value of Hubble's glorious pictures.

The dramatic image of the Eagle Nebula, a gaseous region more than 7,000 light-years away in the constellation Serpens, was pivotal in changing this understanding, and it has become an icon of the Hubble Space Telescope's work. Silhouetted against a velvety blue background, pillars reach up like massive megaliths and glow with an unearthly light, pierced by pink stars. The giant forms fit perfectly in the stair-step frame of the WF/PC 2. NASA and the Space Telescope Science Institute released the image to the public on November 2, 1995. Several major city newspapers featured the image the next day, and newsmagazines ran stories in the following weeks. The articles describe the scene as dramatic, eerie, monstrous, stunning, and breathtakingly beautiful. Because astronomers believed it was a star-forming region, the image soon gained a nickname: The Pillars of Creation.

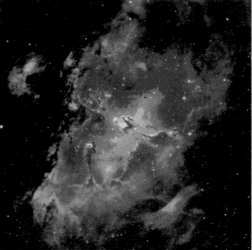

Wide-field mosaic of the Eagle Nebula in Serpens from Kitt Peak National Observatory.

In June 1996, *The Astronomical Journal* published the scientific results, which offered a more detailed analysis of the data gathered by HST. Using the high-resolution pictures, astronomers Jeff Hester and Paul Scowen identified in the gaseous columns smaller structures they called evaporating gaseous globules, or EGGs, which they theorized were the sites of new stars in the process of forming. The EGGs contained stellar material enveloped in a molecular cloud. Over time, radiation from a nearby massive star evaporated the molecular cloud, eventually separating the stellar object from the larger column. If the proto-star was large enough in size, it might then develop into a true star.

The image's high resolution allowed astronomers to see star formation in more detail than possible from ground-based observations, but, arguably, the image's appearance—and the enthusiastic response to it—had as great an impact as its

II.2.95 Star birth in pillars of gas and dust at the center of the Eagle Nebula.

scientific content. As an acknowledgment of the value of the image's visual appeal aside from any scientific conclusions that could be drawn from it, Hester and Scowen closed their article with these words: "[T]he WFPC2 images of M16 are visually striking, and noteworthy from that perspective alone. They also provide a fascinating and enlightening glimpse of the physical processes at work in the interplay of massive stars and their surroundings."[ii]

With the publication of more scientific articles based on the repaired telescope's observations, recognition of the images' aesthetic qualities increased. Astronomers began to use words like "dramatic" and "spectacular" to describe the Hubble images, rather than limiting themselves to the dry, analytical vocabulary typical of scientific articles.[iii]

The appearance of the images released to the public also began to change. Until 1995, Space Telescope Science Institute press releases often featured black and white images. Color might be used, but more often it was a single hue rather than a full-color array. Over the next few years full-color images became much more common. By the fourth servicing mission in 2002, which featured the installation of a more sensitive wide-field camera called the Advanced Camera for Surveys (ACS), the Institute had recognized the public value of the images. Four dramatic full-color views were released to the public: the Cone Nebula, the Tadpole, the Mice, and the Swan Nebula. These images again garnered headlines. An editorial in the *New York Times* concluded: "[The Hubble] has taught us to see properties of the universe humans have been able, for most of their history, to probe only with their thoughts."[iv]

THE CRAFT OF IMAGING

The visual characteristics of the images influence how we see, understand, and respond to these scenes of distant and alien places. For astronomers, focused only on scientific analysis, visual appeal is a secondary concern. The images they create can be anything but beautiful.[v] If an astronomer plans to publish an image in a professional journal, the tools available in IRAF are adequate. Images that are used for press releases, posters, and popular magazines, however, typically undergo another level of image processing, often with well-known graphics programs like Photoshop.[vi] With these,

image processors can use a wider variety of tools to enhance the appearance of an image. From the same data a stunning and spectacular image can be created. Hubble images are mediated views of the universe, and choices made during image processing affect their appearance. In effect, astronomers and image processors employ a degree of artistry to make the universe understandable and attractive. Although derived from scientific data, the images reflect a desire to balance science, aesthetics, and communication.

The incredibly sensitive instruments aboard HST collect a range of data beyond a human's limited visual abilities or even what can be displayed on a computer screen or the printed page. By analyzing numeric data, astronomers can detect the subtle distinctions in light intensity registered by the instrument. To create a representation of these variations that can be sensed by the eye, image specialists will adjust the light intensity scale of the data. Using software, they stretch light values in the middle of the range and allow those at the darkest and brightest ends to be lost in white or black. The resulting image reveals fine details instead of only faint or undetectable distinctions. But it is important to

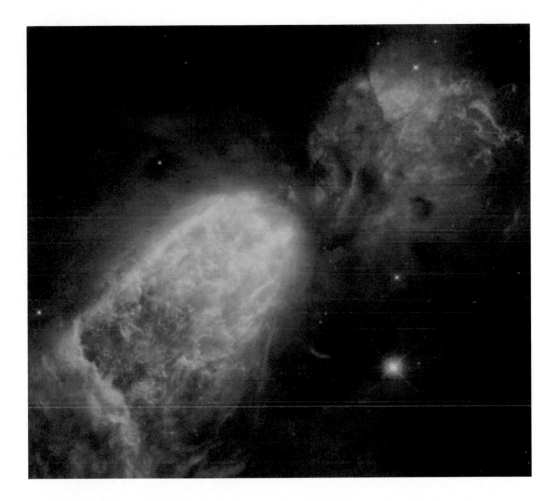

Bipolar cloud of ionized hydrogen in Cygnus regarded as the chrysalis of an infant star.

understand that these adjustments are not determined by a set of universal standards; they depend on the object being observed, the areas of greatest interest, and the judgment of the image specialist.

Once the image processor has maximized the visibility of structure in the image, the next step is to enhance the scene with color. The Hubble Space Telescope returns monochromatic images, although through the use of filters, it can detect the presence of different natural phenomena with certain color characteristics. The full-color images most commonly associated with the Hubble Space Telescope are actually composites of three different images of the same object taken through three different filters. Each image is assigned a different color, usually red, blue, or green, to create a full-color composite image. The physical conditions present in the celestial object being studied often influence

The Landscape of Space

The Art of the Hubble Image

The HST images depict distant skyscapes and alien objects, yet the scenes do not appear wholly unfamiliar; they resemble earthly landscapes. The press release that accompanied the Eagle Nebula image described the formation of the gaseous columns in landscape terms, as "analogous to the formation of towering buttes and spires in the deserts of the American Southwest."[1] The comparison appears again with the Cone Nebula, which the press release portrays as a "craggy-looking mountaintop." But more than just looking like buttes and mountains, the images continue a tradition of representing the awe and majesty of nature. In the last decades of the 19th century, artists such as Frederic Church, Albert Bierstadt, and Thomas Moran created paintings of the American West with a similar sensibility. Traveling through Colorado, Wyoming, Arizona, California, and other western states, these artists provided an eager public back home with a glimpse of the incredible scenery of the already mythical West.

In 1871, Moran accompanied a government-sponsored expedition surveying the Yellowstone region. Under the lead of F.V. Hayden, the group not only created detailed maps, but also collected scientific data; Moran documented the landscape they covered as well. Like the Hubble Heritage Project images, his work can be seen as a federally supported effort to forward the interests of science through art. Purchased by Congress, *The Grand Canōn of the Yellowstone* once hung with another of Moran's paintings, *The Chasm of the Colorado* (1873-74), in the Capitol Building, and now both paintings are in the collection of the Smithsonian American Art Museum.

The HST images and Moran's paintings also share a set of visual characteristics. High contrast between light and dark, vivid colors, and suggestions of vastness and infinity are prominent in both. In *The Grand Canōn of the Yellowstone,*

Thomas Moran's *The Grand Canõn of the Yellowstone*,
an exemplar of art evoking endless yet accessible horizons.

areas of shadow alternate with regions flooded in brilliant light. Although a rock canyon permits only a restricted palette, Moran carefully incorporates vivid hues whenever possible. The intense blue of the river contrasts with the warm yellow tones of the canyon walls, and rich greens frame the lower portion of the vista.

Moran's paintings also suggest scale. Figures dwarfed by the landscapes that spread before them are a measure of the immensity of the scene, but they also promise the possibility of experience. Moran and his peers in the 19th century extended an invitation. New railways, hotels, and other conveniences would soon make the sites of his paintings accessible.

Whether used for science or for public display, all HST images are mediated through the telescope. Still, a human hand is very much at work in communicating the majesty of the images. Instead of painter at the site, astronomers and image processors on the ground use their knowledge and experience to interpret the data and create a compelling representation. The resulting images of nebulae, galaxies, and star fields recall a well-established aesthetic tradition and in the process make these distant spacescapes familiar.

how colors are associated with these filtered views. In other words, image processors determine whether an image taken with one kind of filter should appear as red, or whether it would be more appropriate for it to display as blue or green. If a filter detects the presence of a gas that glows within a certain spectral range, then this characteristic may determine the color used. For example, hydrogen glows red, and when a specific filter in the camera is selected to study it, the resulting image is often displayed as red in the composite. But the conventions used by astronomers to determine which of many filters in WF/PC 2 or ACS are associated with which color are complex. If two distinct processes in the object happen to glow in the same color or spectral range, then the astronomer has to choose another color for one of them to keep them distinct, even if that color has no physical relationship to the filter.

It is also common to associate color with the level of physical excitation of the gas, or its temperature. The image through a filter detecting the hottest gases will typically be assigned to blue, while one registering the coolest gases will be assigned to red. Yet as the existence of two systems for assigning color suggests, color decisions are not governed by strict rules, and the final choice often reflects a mixture of scientific conventions and aesthetic judgments.

In addition to contrast adjustments and color assignments, highly refined images require cosmetic cleaning. IRAF will automatically eliminate some of the noise, instrument artifacts, and cosmic rays. Other blemishes, such as seams between the exposures in the mosaic, are handled in Photoshop, as are bright and dark spots caused by flaws in the camera. Retaining these defects would distract from the subject represented. Yet, if you look carefully, in many images you will find diffraction spikes—thin pointed beams of light caused by obstructions within the telescope—around bright stars. These spikes are familiar in celestial scenery, and also they function as strong visual cues. Image processors therefore keep them in the picture because of their symbolic and aesthetic value.

Other choices may help to transform a dull, disorganized, or uninteresting scene into an arresting one. While astronomical images have traditionally been oriented with north at the top and east to the left, Hubble's images are often oriented to create the greatest visual interest. An image may be cropped to remove distracting or less interesting features.

6.5.03 Portion of the edge of a supernova shockfront, the Pencil Nebula, in Vela.

4.30.02 ACS image of a tidally disrupted galaxy, the Tadpole, in Draco.

It may be combined with data from other sources to add interest or compensate for missing sections of data. The field of view can be modified with the addition or subtraction of data to create a more conventional rectangle or square rather than a strangely shaped image caused by the design of the camera. Any of these adjustments to the image's composition may also be used to achieve a better sense of balance in the picture.

The flexibility of the digital medium and the uncertainty about the roles of science and aesthetics can raise questions about the truthfulness of the images, and some commentators view the Hubble images with suspicion. Understanding the procedures used in creating the images can help to address these concerns. Modifications to contrast and color, which help to enhance detail and structure, alter light intensity in the image. By focusing on one aspect of the data, another aspect might be sacrificed. These adjustments, however, occur at the level of the image, not the data, meaning that if an astronomer sees something interesting in the image, he or she can return to the data to make measurements and calculations.

But more than understanding how the images are processed, the viewer needs to recognize that Hubble images should not be compared to our visual experience of the world. The instruments used in the Hubble Space Telescope do not repeat the behavior and functioning of the human eye. Instead, they vastly expand upon it, registering light at wavelengths beyond our visual range and collecting light over extended periods of time. Given Hubble's exceptional optical capabilities, it would be inaccurate to restrict its images to the limits of human eyesight or to judge them by the same standards of truthfulness humans use for their eyes. These images are not representations or simulations of a person's visual experience, regardless of whether they looked through a telescope or traveled to a distant region of the universe. Instead the final images are impressions, based on scientific data, that show the universe in a visually compelling manner.

We have a responsibility to be savvy viewers and consumers of images. The Hubble Space Telescope images are valid impressions, and astronomers and image processors are not in the business of deceiving or fabricating. They have made con-

certed efforts to document the steps between data and display. But it must be recognized that in the work to clarify and extract information from the data, the image moves further from the actual phenomenon and closer to an abstraction. Critical aspects of the original are preserved, although they are transformed into knowledge that can be understood. The work of science is often described as moving from tangible material to abstract numbers.[vii] Here the direction appears at first to be reversed, moving from abstract numeric data returned by the telescope to images. It is easy to be seduced by images, believing they bring one closer to concrete, perceivable originals; in truth they reflect instead an abstraction made visible.

HUBBLE HERITAGE PROJECT

Creating a defect-free image that exhibits a full range of light and color requires both time and expertise in image processing. The Hubble Heritage Project, an effort staffed and sponsored by the Space Telescope Science Institute, uses these techniques to explore the aesthetic potential of the data collected by the Hubble Space Telescope. The project started in the spring of 1997 after a group of astronomers—Keith Noll, Howard Bond, Anne Kinney, and

10.21.98 The first Hubble Heritage release, clockwise: Saturn, a spiral galaxy, a star cloud in Sagittarius, the Bubble Nebula.

The Hubble Heritage Team

Artists and Scientists Working as One

At the Space Telescope Science Institute, the Hubble Heritage Team members gather. Zoltan Levay, an image specialist who works for the Institute's Office of Public Outreach, has just put the finishing touches on an image and has brought it to the group for approval.

The group has looked at several versions of this image. Keith Noll, the Heritage Project's principal investigator, found a set of three exposures in the HST archive and shared an initial color version. All agreed that the image had potential, but more work was needed to bring out the structure. Two other team members—Lisa Frattare and Forrest Hamilton—started some of the first image-processing steps. The group then looked at several different compositions, keeping in mind that certain astronomers might object to an image that did not maintain the convention of north on top, east to the left. Although they considered this option, in the end they chose an unconventional orientation because they felt it created a more visu-ally dynamic picture. Color selection, the topic of another meeting, followed a more traditional approach; most astronomers would recognize it from images of this object from other telescopes.

Howard Bond wrote a caption to accompany the image, and Carol Christian helped to coordinate the release through the press office; the group could look forward to seeing this new image on the website.

After a similar series of discussions, other images will be ready to add to the Heritage Project's growing gallery. A collaborative effort is required for each image. The team tries to create visually compelling pictures for the public, yet remain sensitive to the responses of their colleagues at the Institute. Although some efforts are coordinated with other public outreach activities, the Heritage Project maintains a level of independence. Positioned at the intersection of art, science, and public relations, the Heritage team attempts to balance the interests of all.

Carol Christian, all of who were working at the Space Telescope Science Institute—recognized the opportunity to create a lasting legacy for Hubble.

Although the Hubble rendition of the Eagle Nebula demonstrated the aesthetic appeal of the images, only a small percentage of the images produced by the telescope rival it. A scientific question might as easily be answered by framing only a small portion of an object rather than the complete form. Multiple observations with different filters, necessary to create a full color image, may not be required to reach a valid scientific conclusion and therefore are not gathered. Astronomers are not specialists in image processing, and limited resources in terms of telescope observing time and astronomers' time also constrain the production of impressive images.

Although aware of these constraints, the Hubble Heritage founders continued to be convinced that people who knew little about the science behind an observation needed to gain an appreciation of the celestial wonders it represents, if for no other reason than to raise high-minded questions about humanity's place in the universe. The Hubble Heritage founders proposed a program that would remain true to the

scientific content of the images, while celebrating their aesthetic power. They presented the idea to the director of the Space Telescope Science Institute, Robert Williams, requesting a small budget and some observing time on the telescope. With his approval, they formed a team composed of astronomers, image processors, and interns who would work collaboratively to generate one visually compelling image each month.

Beginning with a series rather than a single picture, the group released its first images in October 1998. The four images—a planet, a star field, a nebula, and a galaxy—constituted a kind of summary of Hubble's observations. Saturn is depicted against a flawless black sky, its rings tracing a strong diagonal line across the picture plane. The Sagittarius Star Cloud teems with countless stars that shine in brilliant hues. The Bubble Nebula combines both a strong diagonal and vivid colors with two amorphous forms glowing in opposing corners of the image. In the galaxy image, halos of bluish stars surround the brilliant white and yellow core. Keith Noll summarized, "These images communicate, at a visceral level, the awe and excitement that we experience when exploring the universe with

5.31.01 Galaxy image is a composite of seven images produced through different filters and with three Hubble instruments: WF/PC 2, FOC, and NICMOS.

Hubble. It is our chance to repay the public that supports us."[viii]

The Hubble Heritage team obtains data for the images by searching the Institute data archive for visually interesting objects, particularly those with enough exposures through different filters to create a full-color image. The initial quartet of images all came from the archive. In addition, the project requests a small number of orbits from the director's discretionary time. The director of the Space Telescope Science Institute is given ten percent of the observing time to distribute as he determines most appropriate. (Often this discretionary time is used for "targets of opportunity," astronomical events for which no proposal has been submitted, and special projects that require a large quantity of observing time, such as the Deep Field, discussed in the final chapter.) The project's observing time supplements the existing data, such as adding exposures through another filter when only two are available. For example, the Hubble had observed the Ring Nebula through two filters and covered only half of the nebula within the WF/PC 2's frame. With only a few orbits, the Hubble Heritage team collected enough data to show the entire object and create a stunning picture of the well-known celestial form. In order to showcase the capabilities of the new Advanced Camera for Surveys, Hubble was directed to another familiar object, the Sombrero Galaxy, which had no archived data available. Using a much larger number of orbits, the Heritage team created a striking picture months before the public would be able to see the data if gathered through the normal proposal process.

Since beginning the project in 1998, about half the images have come from the archive and half from new observations. Using an average of 25 orbits per year, the Hubble Heritage Project represents less than one percent of the total available observing time. Yet even that is a significant commitment and reflects the importance the Institute and NASA place on creating visually appealing images. Both argue, reflecting Keith Noll's words, that the images repay the public's continued support of Hubble's scientific mission. Rather than rely on fuzzy images, graphs, or text that requires a forbidding level of scientific expertise, the images use a visual language—light, color, and composition—to communicate a sense of wonder and awe for the universe. By using another mode of expression, one perfected within the world of art, the images inspire audiences at every level—from schoolchildren to adults, policymakers to astronomers.

HUBBLE'S AESTHETIC APPEAL

Why do these images lift the human spirit? They depict fascinating objects, but the way in which these objects are presented also plays a role. By skillfully manipulating light, color, composition, and other visual attributes, artists evoke ideas, emotions, and questions in those who see their work. Although derived from scientific data, the images from the Hubble Space Telescope press releases and the Hubble Heritage Project rely on these same artistic techniques to present the universe as awe-inspiring and majestic.

What qualities are desired in Hubble images? They typically exhibit a high contrast between light and dark tones. The Eagle Nebula and the Cone Nebula images include the darkest of blacks in the columns and the brightest whites in the stars and tops of the columns. In the Sombrero Galaxy, the brilliant white of the core contrasts with the dark edge of the dust cloud and the black background of the sky. These examples also contain a full range of tones between light and dark.

Vivid colors emphasize this dynamic range and give the objects a sense of solidity and mass. The pillars in the Eagle Nebula vary from mustard yellows to red, while the background begins as a deep blue at the top of the image and blends to greener tones in the regions surrounding the columns. The intensely glowing red background of the Cone Nebula enhances the three-dimensionality of the form. The image of the Supernova Remnant LMCN49 weaves together blue, gold, red, and green. Against the black and white star field, the cloud of gases becomes almost a tangible form. In another example, the spiral galaxy NGC 1512 appears as a glowing yellow form streaked with red and backlit with blue and purple. While views of galaxies tend to display more subtle tones, color remains an essential element of the images. The greenish arms of the Whirlpool Galaxy are ridged by reddish stars, which highlight the spiral form. These structures twist together into a glowing yellow core.

In the cases of the Eagle and Cone, choices in composition are made to evoke a sense of size. The brightest regions are positioned at the top of the images. This is an arbitrary choice with the Eagle Nebula, since north is located diagonally to the left. These glowing regions pull the eye upward, giving the forms a monumental sensibility. Composition assists in suggesting the vastness of space in other images. The brightest point is often centered, as in the Whirlpool Galaxy or NGC 1512, functioning as a vanishing point would within a landscape painting, drawing the viewer into the image.

This suggestion of movement toward a center is characteristic of a spiral, but it appears in images of other phenomena as well. In the mosaic of the Orion Nebula, the luminosity at the center contrasts with the outer edges and creates a sense of depth. As Christiaan Huygens suggested centuries ago, it appears as if the sky is opening to reveal another realm. A scene of stars and pillars in NGC 3603 follows a similar format. The brilliant blue of the trio of stars at the center of the image draws the eye

10.2.03 ACS mosaic of the Sombrero, an edge-on spiral galaxy in the Virgo cluster of galaxies.

into space, while the glowing tops of the gaseous forms point toward this potentially infinite distance. Centering these phenomena obeys the logic of presenting a specimen for study and observation. It also has an undeniable aesthetic impact for the viewer, suggesting vastness and infinity.

The few images described here and others included in this book have a shared set of attributes: extremes of light and dark, vividness of color. The subject matter certainly merits wonder as the HST brings us new views of the universe, and the appearance of the images elicits this reaction. As a collection, the Hubble Heritage images represent neither the average nor ordinary in nature, but the extraordinary and spectacular.

The images also reflect great interest in the dynamic, even violent, forces of the universe, often portraying colliding galaxies and exploding stars. These same qualities define the experience of the sublime, a notion that was first applied to aesthetics in the 18th century and became associated with experiences of overwhelming grandeur and power, which can elicit feelings of awe, wonder, and transcendence.[ix] Artists throughout the 19th century depicted the landscape as sublime, especially the unexplored frontier. The Hubble images bring us new views of the latest unexplored frontier: outer space. The aesthetic crafting of the images by astronomers and image processors encourages us to respond with awe.

The Harvest

This chapter's gallery showcases the rich and varied imagery created by the Hubble Heritage Team.

2.4.99 Portion of Large Magellanic Cloud centered on supernova 1987A.

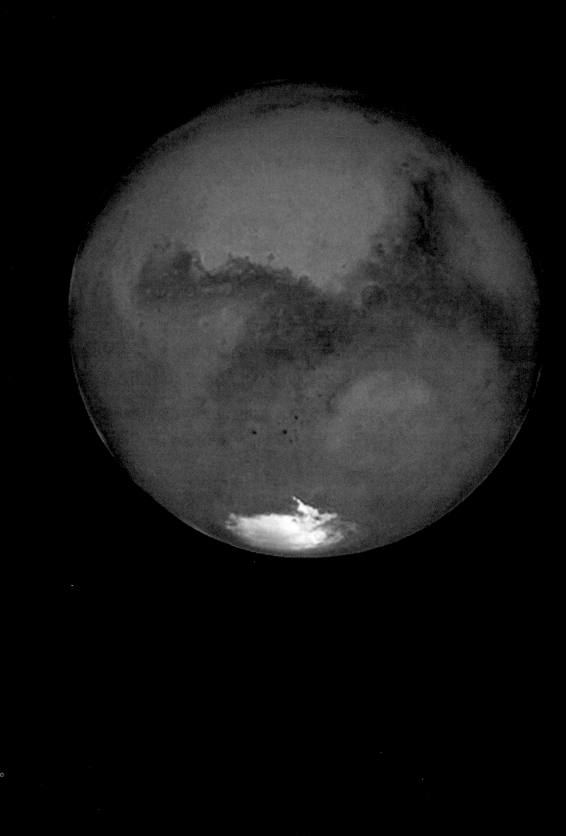

LEFT: 8.27.03 Mars at closest approach to Earth.
Impact features and south polar cap visible.

ABOVE: 8.27.03 Mars' largest known volcano,
Olympus Mons, is the light circle above center.

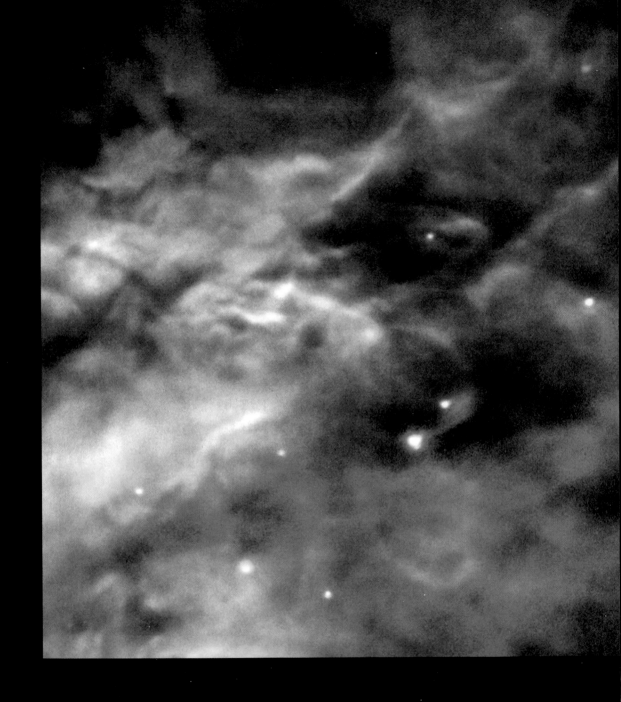

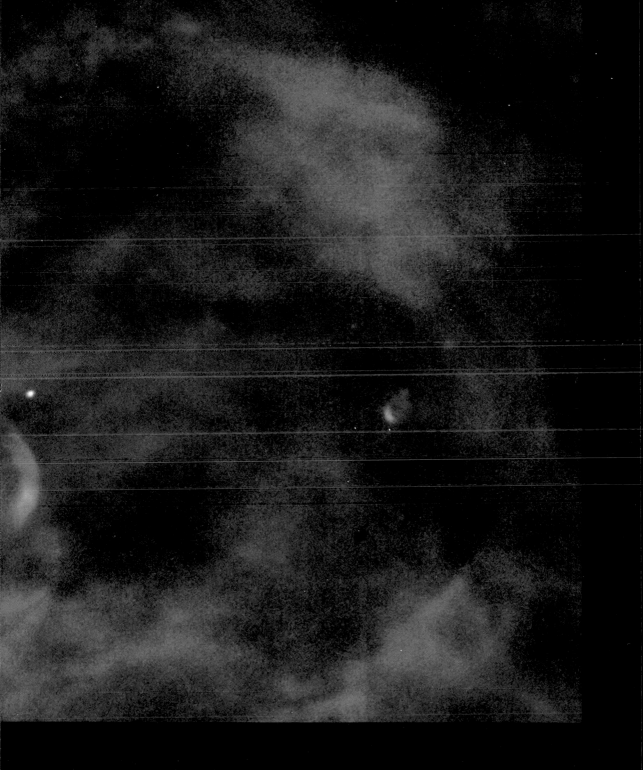

12.16.92 A portion of the Orion Nebula revealing protoplanetary disks, called Proplyds.

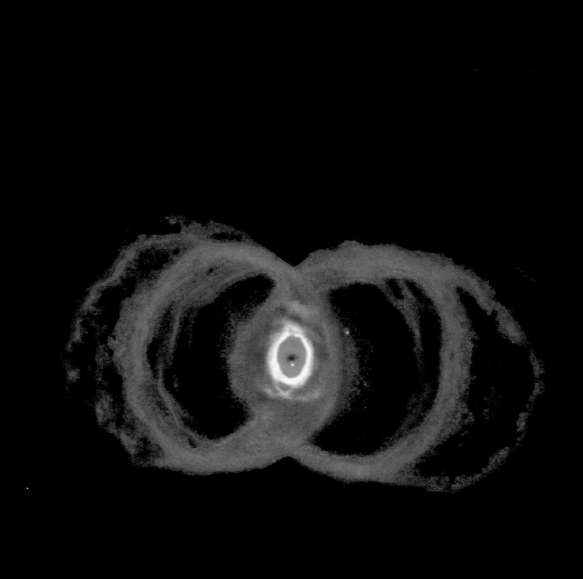

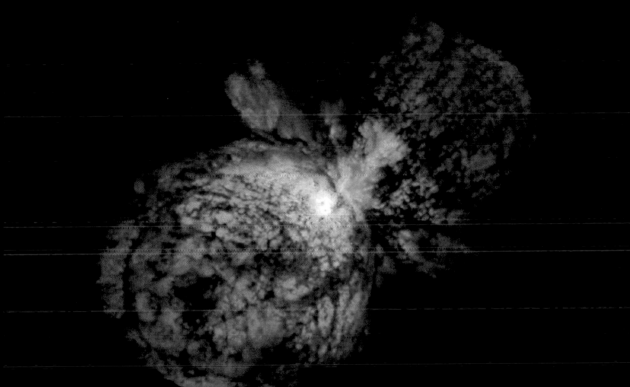

LEFT: 7.16.96 The Hourglass, a young bipolar planetary in Musca.

ABOVE: 6.10.96 Eta Carinae, a star ejecting outer layers and possibly about to explode.

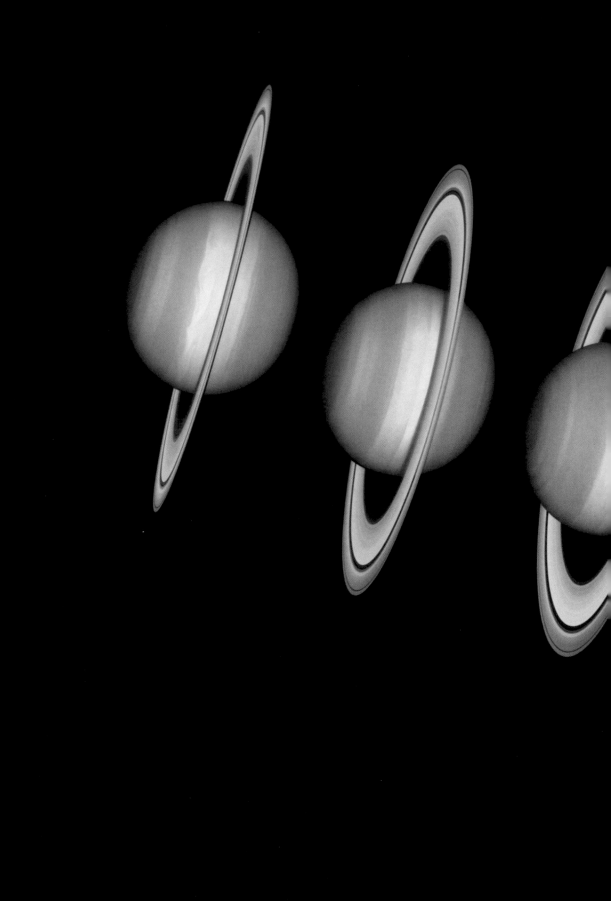

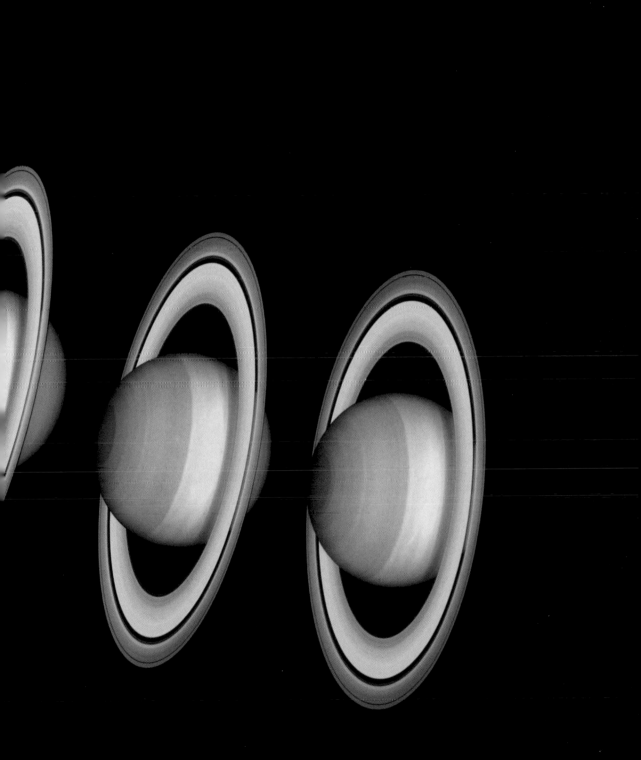

Saturn's changing orientation
between 1996 and 2000.

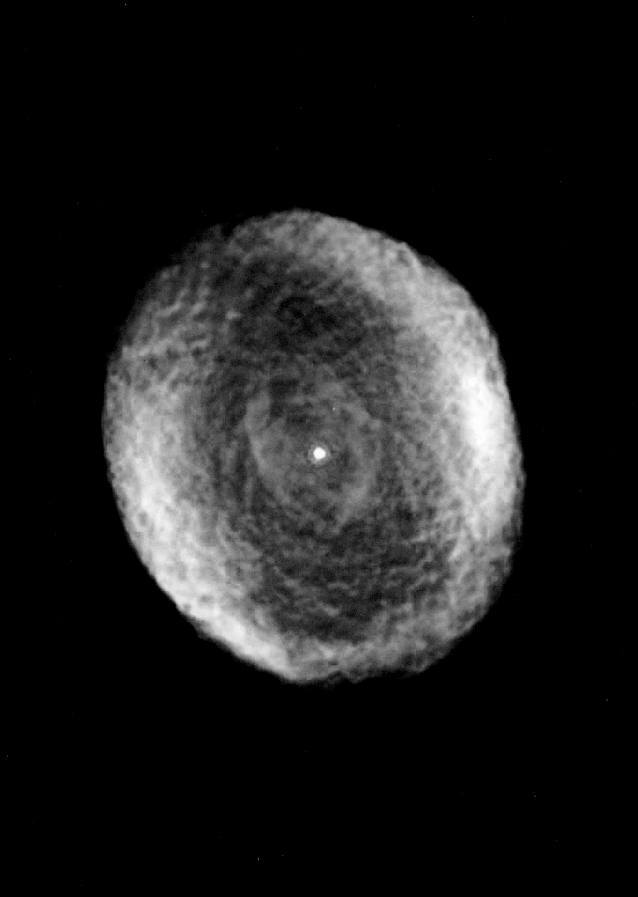

9.7.00 The Spirograph Nebula, a symmetrical planetary in Lepus.

The Harvest

THE DEEP FIELD

THE HUBBLE SPACE TELESCOPE HAS CONFIRMED AND EXPANDED OUR IDEAS ABOUT MANY KINDS OF ASTRONOMICAL OBJECTS AS WELL AS OUR VIEW OF THE UNIVERSE. IN FOUR EXCITING AREAS HUBBLE HAS HAD A MAJOR IMPACT: THE FORMATION AND EVOLUTION OF GALAXIES (THE DEEP FIELD); HOW FAST THE UNIVERSE IS EXPANDING AND HOW OLD IT IS (THE HUBBLE KEY PROJECT); how stars are born, live their lives, and die in spectacular paroxyms of light and mass dispersal; and how the planets of our own solar system can be changed by the impact of cometary debris.

When Robert Williams became the second director of the Space Telescope Science Institute, one of his many tasks was to administer the director's discretionary time. The success of the first shuttle servicing mission in late 1993 induced Williams to convene an international advisory committee of specialists in extragalactic studies to explore how best to use HST during his discretionary block of time. The proposed ideas ranged widely, but eventually all of the members agreed to a deep survey of a carefully selected part of the sky. Whenever a new and more powerful telescope comes on the scene, or a telescope capable of detecting wavelengths of light in unexplored regions of the electromagnetic spectrum, astronomers typically conduct new surveys of the whole sky. In the case of the Hubble, however, a deep, open-ended survey of the entire sky was not feasible. Hubble's small field of view meant such a project would take an impossibly long time. Instead, one of Hubble's great strengths is in imaging tiny features, allowing astronomers, for example, to examine the structure of galaxies a small fraction of the age of our own. Williams's committee therefore settled on one or two long exposures of one area of the sky, both as a scientific mission and a public legacy for HST.[i] In coming to this decision many issues were debated: the number of regions of the sky to examine; where the telescope would point; how many filters would be used; whether it was appropriate for already known objects to be included in the field. Increasing the number of filters improved the possibility of discriminating structure, composition, and other features, but limited the exposure times for the finite number of orbits available to the team. Since everyone wanted HST to penetrate into space as deeply as possible, exposures had to be as long as possible so that the light of extremely faint galaxies would be recorded. Further, the exposures had to be of regions of space that were not blocked or partly blocked by foreground stars, galaxies, or clouds of dust and gas. In addition, the field had to be always visible as Hubble sped around its orbit, not occulted by a planet, the Earth, or the moon.

The 4-meter Mayall telescope at the National Optical Astronomy Observatories main campus at Kitt Peak outside Tucson was chosen to scout for areas of the sky that met the astronomers' requirements. After its search the astronomers decided on a field that lies just above the handle of the Big Dipper, in the far northern sky. Its angular extent is no larger than the appearance of a dime seen from a distance of 75 feet, yet it covers an immense expanse of space, wholly beyond human comprehension. The Deep Field, as the resulting image became known, thus became a core sample of the universe, a composition generated from 342 separate exposures using WF/PC 2 for ten consecutive days in late December 1995.

At the conclusion, traces of 1,500 galaxies in an unimaginable volume of space were captured in a two-dimensional image. One of the astronomers' subsequent tasks was to secure the distances to the galaxies so as to sort out the scene into the eons of time from the present back into the incredibly remote past—75 percent back to the beginning of the universe for the most distant objects.

The Deep Field was presented to the public in January 1996. It sparked a worldwide effort by astronomers to examine this small patch of sky in any way that might

1.15.96 The Deep Field (in the Big Dipper), the apparent size of a dime seen from 75 feet.

add more information to these findings. Within a short time the Deep Field and the objects within it had been probed in the infrared, x-ray, and radio range, by telescopes on the ground and instruments in space. Their spectra were scrutinized by the world's largest ground-based telescopes to determine their distances.

Eighteen months later, astronomers had determined enough distances of objects in the Deep Field to begin to add a third dimension to the initial two-dimensional image. During this process astronomers came to realize that galaxies do not form in huge assemblies that then break into individual galaxies. According to the new view, galaxies actually start out life as small irregular clumps that gather into massive bodies by collision and accretion.

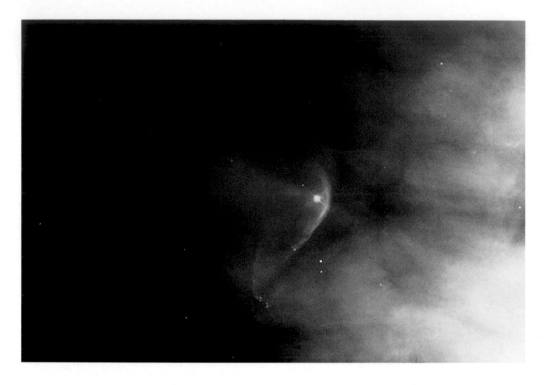

3.6.02 Shock waves around a young star in Orion produced by fierce stellar winds.

Galaxies are cannibals that grow larger by gobbling up other galaxies, and astronomers now see this as the primary way that galaxies evolve.

With the addition of the Advanced Camera for Surveys on the fourth servicing mission in 2002, Hubble was equipped with a camera with a wider field of view than WF/PC 2's, as well as significantly enhanced image quality and sensitivity. ACS has penetrated even deeper into space than the Deep Field through another series of very long exposures called the Ultra Deep Field that have consumed 412 orbits and are devoted to trying to reach an epoch when the very first galaxies and

quasars began reheating the intergalactic medium.

THE HUBBLE KEY PROJECT

When Edwin Hubble first correlated the red shifts of galaxies with their distances in the late 1920s, he reckoned that for every megaparsec one moved out into space, the red shift increased by approximately 500 kilometers per second. Calculations of this term—later known as the Hubble Constant—fluctuated somewhat through the 1930s as Hubble refined his observations. As a result of new measurements of the distances to galaxies, in the decades that followed other astronomers argued that the value was lower than Hubble's initial estimate. By the early 1970s, when planning for HST was seriously under way, quoted values of the Hubble Constant had dropped to 100 kilometers per second per megaparsec or lower.

Astronomers today use the relationship between red shift and distance to estimate distances in the deep universe, required for projects such as the Deep Field, and also to estimate the age of the universe. The rate of expansion is an indication of its age, or, as nearly all astronomers believe, the time since the big bang. The larger the Hubble Constant, the shorter the time since the big bang. Each reduction in the value of the Hubble Constant since the 1930s has therefore meant a lengthening of the estimated age of the universe.

Early estimates of this age based on the Hubble Constant were two billion years—that is, shorter than the calculated ages of many stars. Something was amiss, since this implied the age of the universe was shorter than the ages of many objects within it! Even measurements of the Hubble Constant around 100 gave uncomfortably short ages for the universe. As a result, astronomers wanted to better pin down this fundamental measurement, to make the estimated age of the universe and the estimated ages of stars work in better agreement. When in the late 1970s NASA invited astronomers to build Hubble's scientific instruments, the space agency listed the measurement of the Hubble Constant as the priority problem for the telescope to tackle.

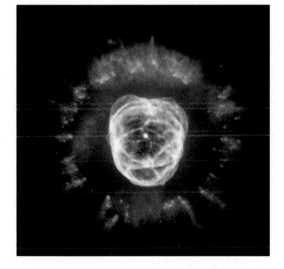

I.24.00 The Eskimo Nebula in Gemini, a complex planetary about 10,000 years old.

To meet this priority, astronomers argued that HST had to be able to detect Cepheid variable stars at least as far away as the Virgo Cluster of galaxies, 20 megaparsecs distant. Cepheids, they believed, were the critical indicators of distance. Determining the Hubble Constant became a key project for HST, and more than 400 hours of observing time was allotted within its first decade. A team of astronomers formed

The Hubble Constant

Calculating the Expanding Universe

In the mid-1920s, no astronomer believed the universe was expanding. By the mid-1930s, almost all did. This rapid change was brought by two developments: the observations of distant galaxies by Edwin Hubble and Milton Humason at the Mount Wilson Observatory and the application of Einstein's theory of general relativity to the universe. Astronomers decided that the rate at which the expansion velocity changes with distance was a crucial measurement for understanding the universe. By the 1950s, this measure had become generally known as Hubble's Constant, H. It is measured in terms of kilometers per second per megaparsec (a megaparsec is one million parsecs, about 3,260,000 light-years): If H has a value of 60 kilometers per second per megaparsec, a galaxy at a distance of 1 megaparsec will be moving away at 60 kilometers per second. The inverse of the Hubble Constant gives a measure of what is often referred to as the age of the universe but is more accurately described as the expansion age. This is not necessarily the same as the time since the universe started

expanding. If, for example, the expansion now observed in telescopes was faster in the past and has decreased to the current rate because of the mutual gravitational attraction of galaxies, the expansion age will be the maximum time passed since the beginning of expansion. If the universe is evolving, as astronomers generally believe, the Hubble Constant will change, and H is better referred to as H_0—the value of the Hubble Constant at present.

If we take our current view of the universe and run time backward, in the distant past all of the space in the universe and all of the energy and matter it contains must have filled a tiny volume. For some astronomers in the 1930s this was absurd. Surely something was wrong in the reasoning. For Georges Édouard Lemaître, the implication was that the universe must have started expanding from such a fantastically dense and hot state and that the expansion was the product of what we call the big bang. Not until the mid-1960s did the idea of the big bang become dominant. Unexpected evidence was secured by

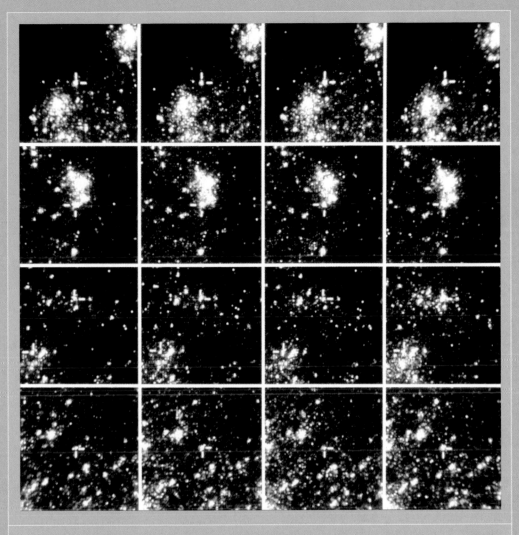

6.9.93 Multiple exposures of four fields in the spiral galaxy M81.
Each field contains a Cepheid variable star marked with white bars.

two radio astronomers in New Jersey. Arno Penzias and Robert Wilson detected a faint hiss of radio noise— traces of the big bang event in the form of very faint microwave radiation. They won the Nobel Prize in 1978 for their research; they had detected the predicted radiation, clinching evidence for the big bang.

When HST was launched in 1990, estimates of the constant, and so the time since the big bang, varied widely, and an accurate determination was judged to be one of the telescope's crucial tasks. HST observations have not settled debate over the Constant, but have constrained its limits to between 58 and 72 kilometers per second per million parsecs, setting an age for the universe of 13 billion to 14 billion years.

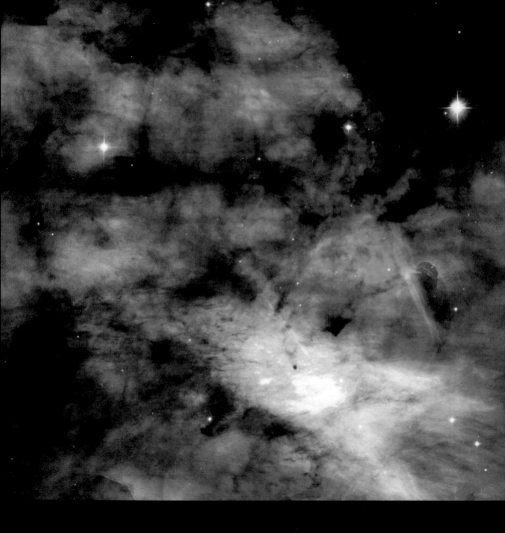

4.30.02 ACS view of the central portion of the Omega Nebula in Sagittarius.

to pursue the project, choosing galaxies to observe likely to contain large numbers of Cepheid stars. Their collective goal was to narrow the constant to within ten percent of its value. At the time, astronomers often joked about not knowing this fundamental constant to better than 50 percent of its value, calling it the "Hubble variable constant."

In their final report in 2001, the 16-member Hubble Constant team, led by Carnegie Institution of Washington astronomer Wendy Freedman, reported that they had reached a value of 72 ± 8 km/sec/mpc, making the universe's expansion age around 13 billion years. However, another team of astronomers, this one led by Allan Sandage, a one-time student of Hubble's, had found a different value of the constant, using a type of supernova to measure distances to far-off galaxies. Sandage's group estimated that the Hubble Constant was closer to 58 ± 5 km/sec/mpc. More recently other estimates have lowered that value into the mid- and even low 50s.

One of the ways astronomers gauge progress in understanding a phenomenon is to find independent ways to assess its properties, and then see if those different routes lead to similar answers. So, when recently an entirely new and powerful method of estimating the age of the universe became available, it was used to assess the Hubble Constant as well. The novel method used microwave studies of the primordial background radiation, what astronomers interpret as the relic of the big bang. The value determined by this independent technique employed aboard a spacecraft known as the Wilkinson Microwave Anisotropy Probe (WMAP) was that the value of the Hubble Constant is 72 ± 0.05 km/sec/mpc, very close to the value produced by the Key Project team led by Wendy Freedman, yielding an age for the universe of 13 billion years. In sum, the results of the Key Project team, the Sandage team and the WMAP study have not settled the question of the value for the Hubble Constant, but at least the spread has been narrowed from 50 to 100 down to a smaller range between 58 and 72.

THE BIRTH AND DEATH OF STARS

In addition to probing the farthest reaches of the universe and its rate of expansion, HST has expanded our knowledge of the various critical stages of the lives of stellar objects, from birth to death. The idea that stars actually have lives—that they are born, live, and die as finite energy-generating and -consuming bodies— did not take a recognizably modern

Albert Einstein and Edwin Hubble (far right) tour Mount Wilson in 1931.

form until the mid-19th century. The big question was: What makes the stars shine for millions of years? In the late 1930s, physicists argued that nuclear energy could do the job, specifically the fusion of hydrogen into heavier elements, carrying with it an enormous release of energy. This was the key to unlocking the secret of the stars. By the 1950s, sources of nuclear energy had been firmly incorporated into theories of stellar evolution; stars were now reckoned to be great globes of glowing gas in balance. The inward pull of gravity, caused by a star's enormous mass, was balanced by the outward push of radiation from thermal and nuclear sources.

Astronomers believe stars like the sun generally form individually or in clusters out of vast clouds of gas and dust. The Orion Nebula and its environs is a fine local example of a stellar nursery, as is the Eagle Nebula. The beautiful high-resolution, full-color Hubble images of these regions actually reveal newborn stars just emerging from their rapidly evaporating sacks of gas and dust. Typically these clouds are evaporated when the most massive, and therefore most energetic, stars start shining, flooding their neighborhoods with high-energy radiation. Spikes and columns of dense material can be seen standing in the shadows of stars illuminated by these superhot neighbors in some images, revealing the erosive effect of such high-energy stars.

The brightest stars in the Pleiades, a young cluster of stars, have driven off most of their crysallis, the dust and gas that originally surrounded them. In contrast, older star clusters are devoid of gas and dust. In observing a number of young stars in some star-forming regions, as well as mature isolated stars, Hubble has also detected disks around them that will in time form into planetary systems. These finds, together with similar ones by other telescopes, have led astronomers to suggest that such disks of material—made up of ice, gas, and dust—are common to stars, thus vastly increasing the likelihood of other solar systems.

Our own sun is reckoned to be far older than the stars of the Pleiades, and its planetary system formed billions of years ago. Stars similar to the sun are very long-lived, typically stable for billions of years. But once the

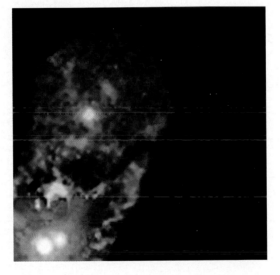

9.21.00 A newborn binary star system in Taurus expels a huge plume of gas.

hydrogen fuel available for fusion in such a star's central regions is exhausted, a series of developments propel the star into becoming a type known as a red giant. When our sun reaches this point in its life it will inexorably expand. It will change its color to red and engulf the orbits of the inner planets. Its outer regions will likely reach beyond even the Earth. Such red giant stars that we see today are necessarily very old stars, so we find them typically in older regions of the galaxy, like in globular clusters.

Red giant stars of the same mass as the sun do not last terribly long by cosmic standards, probably a few hundred million years. Typically, for stars from one solar mass—that is, a star containing as much mass as our sun—to up to eight solar masses, the outer atmosphere dissipates in time. Perhaps it forms what is known as a planetary nebula. These objects have nothing to do with planets, in fact, but William Herschel gave them this name because they presented a

180   120   60

- Solar System
- Stars
- Star Clusters
- ISM/Nebulae
- Galaxies/AGN
- Galaxy Clusters
- Other

Sky map centered on the Milky Way
showing Hubble observations.

disklike appearance in his telescopes at the end of the 18th century, not the pointlike appearance of a star. HST has underlined that they are extraordinarily beautiful and complex objects. With the formation of a planetary nebula, the mass-losing star eventually forms what is known as a white dwarf star, which is hot and extremely dense.

Stars yet more dense than white dwarfs can result from cataclysmic stellar explosions known as supernovae, which produce the most violent events in the universe. Astronomers believe that stars explode for a number of reasons, but true supernovae occur only in stars of at least eight solar masses. These stars exhaust their stores of hydrogen fuel very quickly and rapidly expand to become giants, but there is sufficient mass to hold the star together when the burning of helium, and later other elements such as carbon, starts. But after all the available fuel is exhausted, the core of the star collapses rapidly, making the entire

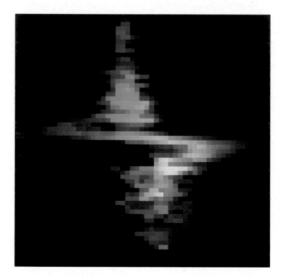

5.12.97 Spectral signature indicating the presence of a supermassive black hole.

star violently unstable. A prodigious amount of energy is released in an incredibly short time, causing a shock wave to travel through the star that literally blows it apart. Most of its mass is spewed outward. As the shock wave penetrates into space, it encounters gas and dust and quickly heats the entire volume, creating a fantastic series of arcs, shells, and explosive remains. The remnant from Supernova 1987A, which exploded in the Large Magellanic Cloud in February 1987, has been investigated closely by Hubble. An older remnant is the Crab Nebula in Taurus, from a supernova explosion less than a thousand years old. More ancient still is the 15,000-year-old Cygnus Loop within the huge Veil Nebula, showing highly dissipated shock fronts.

Supernovae typically result in incredibly compressed cores, denser by far than white dwarf stars, and are called neutron stars. The reason for this is that the material has collapsed in on itself at the atomic level, rendering protons and electrons no longer distinguishable and causing the entire star to behave as a single gigantic neutron. Neutron stars spin in-creasingly swiftly as they are reduced in size, just as an ice skater spins faster as she draws in her arms. If these rapidly rotating pulsars, as they are called, are massive enough, they will continue to collapse to become black holes. These objects are so dense that not even light, which travels at 186,000 miles per second, can escape their gravitational whorl. Astronomers believe enormously massive black holes lurk at the center of many galaxies and power incredibly energetic objects known as quasars. Black holes cannot be seen directly, but reveal themselves by their gravitational presence as they perturb material in their vicinity.

OUR NEAREST NEIGHBORS

In the early planning for what became the HST, the primary goal of the telescope was extragalactic astronomy. But HST was also regarded as an extremely valuable tool for examining planets in our solar system as well as other astronomical targets much closer to home. And so it has proved, with observing time devoted to solar sys-

tem research in each research cycle. The WF/PC 1 and 2 were designed in part to take high-resolution images of planetary surfaces and their attending moons. Some of the other Hubble instruments have examined planetary atmospheres and other properties of comets. Although even Hubble's resolution cannot match what would be possible from a spacecraft sent into orbit around, or one that landed on, a planet, it does enable the long-term monitoring of ever-changing weather patterns of the planets. Hubble's images of Mars, Saturn, and Jupiter are stunning, of course, but they also provide detailed information, and alert astronomers to new features. Perhaps the most remarkable planetary images taken by HST came when, over several days in 1994, Comet Shoemaker-Levy smashed into Jupiter.

As this fragmented comet rushed towards Jupiter, astronomers worldwide joined to observe it with HST and ground-based telescopes. The collisions actually occurred on the side of Jupiter turned away from the Earth, but within a few hours the huge atmospheric shock waves from these crashes were readily detected. Some persisted for over a month and were tracked by HST to provide astronomers with data on the

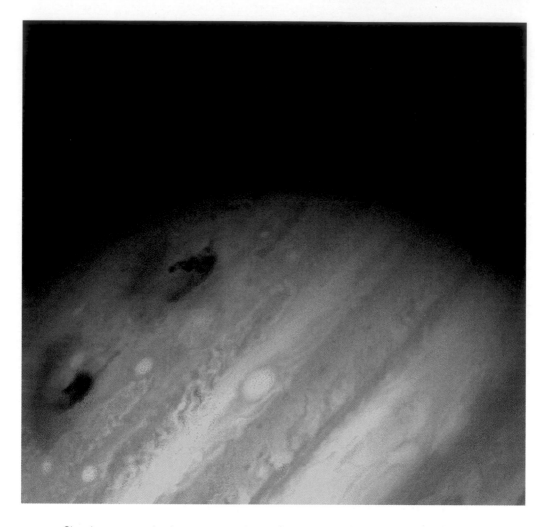

7.22.94 Shock waves in the Jovian atmosphere after impact of the Shoemaker-Levy 9 Comet.

movement of winds in the atmosphere of Jupiter.

Our reconnaissance of Hubble's visual universe thus ends back home, among the planets and other debris that make up our solar system. With Hubble we have traveled multiple billions of light-years and through eons of time, back to when the first galaxies began forming, accreting and clustering. We have explored Hubble's power of "pene-trating into space," reliving the dreams of astronomers of the past half millennium, of Galileo, the Herschels and Rosse, of Hale and Hubble, to reach ever farther into space and to see incredibly distant objects with unprecedented clarity. By sending a telescope into space, we have through Hubble's eyes profoundly changed our view of the universe and the remarkable objects we find in it.

I

N 1987, THREE YEARS BEFORE HUBBLE WAS LOFTED INTO ORBIT, ONE SPACE SCI-ENTIST GAZED INTO HIS CRYSTAL BALL AND SAID, "BY THE TIME ITS TASK IS DONE, PROBABLY SOME THREE DECADES OR MORE HENCE, THE HUBBLE SPACE TELESCOPE WILL ALMOST CERTAINLY HAVE CONTRIBUTED MORE TO OUR UNDER-STANDING OF THE COSMOS THAN ALL PREVIOUS TELESCOPES COMBINED."[i] WHILE THIS

judgment is an exaggeration, without question Hubble's influence on astronomy has been immense.

For nearly four decades in the 20th century a single telescope dominated the astronomical landscape. The 200-inch reflector on Palomar Mountain went into operation in 1948, and for years it was indeed the King of the Hill. By the time HST was launched into space, there were many more astronomers and a much wider range of telescopes. The investigations conducted with Hubble have been extremely important, but they have been both complemented by, and given direction to, researches by numerous other instruments. Once HST had viewed the Deep Field, for instance, many telescopes were then aimed in its direction.

That HST would play such a prominent role in the past 15 years was, however, far from certain. When examining a line of scientific development, there often seems to be an inexorable set of changes at work. But Hubble's history did not unfold in such a manner. At times it seemed the project might founder even

before it got off the ground. The construction of HST saw much inspired engineering, but it was also fraught with difficulties and led to the fiasco of the primary mirror's spherical aberration. That such problems were overcome and that HST has indeed produced superb science is a tribute to the efforts of the thousands of people and hundreds of companies and other institutions across North America and Europe engaged in planning for, building, and operating the telescope.

HST has helped not just to remake the discipline of astronomy, however. It has also reshaped, in a profound way, how both astronomers and the general public view the universe. Although before its launch Hubble's astronomers fully expected its remarkable spatial resolution to produce dazzling images, even astronomers were generally surprised by the beauty and power of the images being generated. Even familiar astronomical objects, when viewed at Hubble's very high resolution, have often shown new and unexpected aspects. As in the middle

12.2-13.93 HST awaiting repairs in the payload bay of the shuttle *Endeavour*.

of the last century photographs from the 200-inch telescope set the level for astronomical textbooks, so Hubble images have set the current standard.

The Hubble Space Telescope, as an example of the biggest sort of Big Science, has been such a costly project that it had to be justified to the government not just once, but over and over again. The images created from Hubble data have certainly helped in that process. But that is only part of the story. What had not been understood before launch was the startling extent to which news about HST and its images would grab the attention of newspapers, magazines, and television, as well as being widely disseminated via the Internet and books.

HST's images have traveled far beyond the pages of scientific journals and textbooks. Some have become cultural icons. They have graced the cover of *Time* magazine. They have been featured on prime-time news. They have even appeared in advertisements for electronics hardware and software, adorned T-shirts, wristwatch faces, coffee mugs, calendars, and CD covers and been the subject of countless popular writings. A collection of 14 of the images are accessible to blind and visually impaired people through use of a special printer that embosses the images on the pages of a book so readers can run their fingers along the contours. HST and the vistas it has opened have clearly struck a deep chord.

Hubble has already assumed a very prominent place in the history of constructing bigger and more powerful telescopes. Novel telescopes like HST spring in part from a desire to address specific scientific problems, but also from an abiding faith that a telescope more powerful in various ways than its predecessors will make possible major new finds and so reshape our views on the universe. As Lyman Spitzer put it in 1946 in his report on the "Astronomical Advantages of an Extraterrestrial Observatory," the most significant discoveries to be made with a large telescope in space

would be those astronomers never expected. He also reckoned that "the chief contribution of such a radically new and more powerful instrument would be, not to supplement our present ideas of the universe we live in, but rather to uncover new phenomena not yet imagined, and perhaps modify profoundly our basic concepts of space and time."[ii]

In January 2004, NASA announced that in the light of safety concerns after the loss of the shuttle *Columbia* in 2003 and the deaths of its seven astronauts, the next planned servicing mission to Hubble would be canceled. In fact, NASA decided not to send any more servicing missions to Hubble. Some critics of the decision thought that Hubble's yearly budget, as well as safety, was at issue. But if indeed there are no more space shuttle missions to Hubble, it spells the telescope's death, likely within a period of four or five years as various components fail and cannot be replaced. Particularly critical in this respect are the gyroscopes used to point the telescope to different regions of the sky, which are relatively short-lived. Nevertheless, Hubble should have at least a few more years to further expand the grasp of humankind and underline its reputation as one of the most remarkable and influential scientific tools ever built.

Endnotes, Suggested Reading, Credits, Acknowledgments, Index

CHAPTER ONE

[i]Owen Gingerich and Albert van Helden, "From Occhiale to Printed Page: The Making of Galileo's Sidereus—Nuncius." *Journal for the History of Astronomy* 34, 2003, pp. 251-267, on p. 251.

[ii]Albert Van Helden, "The Invention of the Telescope." *Transactions of the American Philosophical Society* 67 pt. 4, 1977.

[iii]Mary G. Winkler and Albert Van Helden, "Representing the Heavens: Galileo and Visual Astronomy," *Isis* 83, 1992, pp. 195-217; on p. 195.

[iv]Michael Hoskin, *The Cambridge Illustrated History of Astronomy.* Cambridge; New York: Cambridge University Press, 1997.

[v]Simon Schaffer, "The Leviathan of Parsonstown: Literary Technology and Scientific Representation," in Tim Lenoir, ed., *Inscribing Science: Scientific Texts and the Materiality of Communication.* Palo Alto: Stanford University Press, 1998, 182-222, on p. 203.

[vi]Schaffer 1998, pp. 203-204.

[vii]William Huggins, *The Scientific Papers of Sir William Huggins.* London: William Wesley, 1909. p. 114.

[viii]Alex Pang, "Technology, Aesthetics, and the Development of Astrophotography at the Lick Observatory," in Tim Lenoir, ed., *Inscribing Science: Scientific Texts and the Materiality of Communication.* Palo Alto: Stanford University Press, 1998, 223-248, on 224.

[ix] Pang, 1998, p. 223.

[x]Helen Wright, *Explorer of the Universe: A Biography of George Ellery Hale.* 1966. Repr American Institute of Physics, 1994.

[xi] Wright, *Explorer,* p. 320.

[xii]Robert W. Smith, *The Expanding Universe: Astronomy's "Great Debate" 1900-1931.* Cambridge: Cambridge University Press, 1982.

CHAPTER TWO

[i]H.N. Russell, "Where Astronomers Go When They Die," *Scientific American,* 149(1933), 112-3.

[ii]A number of the historical issues discussed in this and the next chapter are examined at greater length in Robert W. Smith (with contributions by P. Hanle, R. Kargon, and J. Tatarewicz), *The Space Telescope: A Study of NASA, Science, Technology and Politics,* Expanded Paperback Edition, New York: Cambridge University Press, 1993.

CHAPTER THREE

[i]Oral History Interview, J. Rosendahl with J. Tatarewicz, September 12, 1984, National Air and Space Museum Collection.

[ii]Oral History Interview, J. Welch with R. Smith and P. Hanle, August 20, 1984, National Air and Space Museum Collection.

[iii]E-mail, Joe Dolan to High Speed Photometer Team, May 21, 1990.

[iv]Oral History Interview, S. Faber with R. Smith, June 3, 1992, National Air and Space Museum Collection, and S. Faber Notebook Entry for June 19, 1990.

[v]John Noble Willford, "First Hubble Findings Bring Delight," *New York Times,* August 30, 1990, p. B 10.

[vi]*Time,* January 4, 1993, p. 60.

[vii]Steve Maran, quoted in Jim Detjen, "Scientists Come Close to Age of Universe," *Philadelphia Enquirer,* January 17, 1991, p. 18.

CHAPTER FOUR

[i]N. Z. Scoville, M. Poletta, S. Ewald, S. R. Stolovy, R. Thompson and M. Rieke, "High-Mass OB Star Formation in M51: Hubble Space Telescope H $\propto$ and Pa $\propto$ Imaging," *The Astronomical Journal* 123 (2001) 3017-3045, on p. 3018.

[ii]Scoville et al 2001, p. 3018.

[iii]Ray Villard, "From Idea to Observation: The Space Telescope at Work," *Astronomy*, June 1989, 38–44, on p. 43.

[iv]Scoville, et al, 2001.

Chapter Five

[i]This chapter is written by Elizabeth A. Kessler and based on her dissertation, "Spacescapes: Aesthetics and the Hubble Space Telescope Images," University of Chicago, forthcoming.

[ii]Paul Scowen, et. al., "Hubble Space Telescope WFPC2 Imaging of M16: Photoevaporation and Emerging Young Stellar Objects," *The Astronomical Journal* 111 (1996), pp. 2349-2360, on p. 2360.

[iii]Jon A. Morse et al. "Hubble Space Telescope Wide Field Planetary Camera 2 Observations of Eta Carinae," *The Astronomical Journal* 116 (1998) pp. 2443-2461, on p. 2444. And C.R. O'Dell and Shui Kwan Wong, "Hubble Space Telescope Mapping of the Orion Nebula I. A Survey of Stars and Compact Objects," *The Astronomical Journal* 111 (1996) 846-855, on p. 848.

[iv]"The Hubble Achievement," editorial, *New York Times*, 3 May, 2002, late ed., sec. A: 22.

[v]Michael Lynch, and Samuel Y. Edgerton, Jr., "Aesthetics and Digital Image Processing: Representational Craft in Contemporary Astronomy," In *Picture Power: Visual Depiction and Social Relations* [Sociological Review Monography 35] London: Routledge, 1988.

[vi]Ray Villard and Zoltan Levay, "Creating Hubble's Technicolor Universe," *Sky and Telescope* Sept. 2002: 28-34.

[vii]Bruno Latour, *Pandora's Hope: Essays on the Reality of Science Studies*, Cambridge: Harvard University Press, 1999.

[viii]"Astronomers Unveil Colorful Hubble Photo Gallery," Space Telescope Science Institute press release, October 21, 1998. Hubble Site http://hubblesite.org/newscenter/archive/1998/28/text.

[ix]Edmund Burke, *A Philosophical Enquiry into the Origin of Our Ideas of the Sublime and the Beautiful*, ed. Adam Phillips, Oxford: Oxford University Press, 1998.

[1]"Hubble's New Camera Delivers Breathtaking Views of the Universe." *Hubble Space Telescope News*, April 30, 2002. Space Telescope Science Institute.
http://sites.stsci.edu/pubinfo/PR/2002/11/pr.html

Chapter Six

[i]M. Livio, S. M. Fall and P. Madeu, eds. *The Hubble Deep Field*. New York: Cambridge University Press, 1998, on p. 27.

Epilogue

[i]David H. Clark, *The Cosmos from Space*. New York: Crown Publishers, 1987, on p. 126.

[ii]Lyman Spitzer, Jr., "Astronomical Advantages of an Extra-Terrestrial Observatory," Douglas Aircraft Company, Sept. 1, 1946.

Suggested Reading

Berendzen, Richard, Richard Hart, and Daniel Seeley. *Man Discovers the Galaxies*. New York: Science History Publications, 1976.

Christianson, Gale E. *Edwin Hubble: Mariner of the Nebulae*. New York: Farrar, Straus, Giroux, 1995.

Clair, Jean, ed. *Cosmos: From Romanticism to the Avant-Garde*. New York: Prestel Publishing, 1999.

Coles, Peter, ed. *The Icon Dictionary of the New Cosmology*. Cambridge, Mass.: Icon Books, 1999.

Crowe, Michael J., *Modern Theories of the Universe. From Herschel to Hubble*. New York. Dover, 1994.

DeVorkin, David H., ed. *Beyond Earth: Mapping the Universe*. Washington, D.C.: National Geographic, 2002.

Galileo, *Siderius Nuncius or The Sidereal Messenger*, translated with introduction, conclusion, and notes by Albert Van Helden. Chicago: University of Chicago Press, 1989.

Gingerich, Owen. *The Book Nobody Read: Chasing the Revolutions of Nicolaus Copernicus*. New York: Walker & Company, 2004.

Goodwin, Simon. *Hubble's Universe: A Portrait of Our Cosmos*. New York: Viking Penguin, 1997.

Gribbin, John, *The Birth of Time: How Astronomers Measured the Age of the Universe*. New Haven, Conn.: Yale University Press, 1999.

Hetherington, Norriss S., ed. *Encyclopedia of Cosmology: Historical, Philosophical, and Scientific Foundations of Modern Cosmology*. New York: Garland Publishers, 1993.

Hoskin, Michael A. *William Herschel and the Construction of the Heavens*, with astrophysical notes by D.W. Dewhirst (1st American ed.). New York: W.W. Norton & Company, 1964.

Hoskin, Michael A., ed. *The Cambridge Illustrated History of Astronomy*. New York: Cambridge University Press, 1997.

Hubble, Edwin. *The Realm of the Nebulae*, with a foreword by James E. Gunn. New Haven, Conn.: Yale University Press, 1982.

King, Henry C. *The History of the Telescope*, with a foreword by Sir Harold Spencer Jones. Cambridge, Mass.: Sky Publishing Corporation. 1955.

Learner, Richard. *Astronomy Through the Telescope*. New York: Van Nostrand Reinhold, 1981.

Livio, Mario, Keith Noll, Massimo Stiavelli, and Michael Fall, eds. *A Decade of Hubble Space Telescope Science* (proceedings of a conference at the Space Telescope Science Institute, held in Baltimore, Md., April 2000). New York: Cambridge University Press, 2003.

Overbye, Dennis. *Lonely Hearts of the Cosmos: The Story of the Scientific Quest for the Secret of the Universe.* Boston: Back Bay Books, 1999.

Panek, Richard, *Seeing and Believing: How the Telescope Opened Our Eyes and Minds to the Heavens.* New York: Viking, 1998.

Petersen, Carolyn Collins, and John C. Brandt. *Hubble Vision: Further Adventures with the Hubble Space Telescope.* New York: Cambridge University Press, 1998.

Petersen, Carolyn Collins, and John C. Brandt, *Visions of the Cosmos.* New York: Cambridge University Press, 2003.

Preston, Richard. *First Light: The Search for the Edge of the Universe* (1st rev. ed.). New York: Random House, 1996.

Rector, Travis, et al. "Digital Image Processing Techniques to Create Attractive Astronomical Images from Research Data." *Astronomical Journal* (forthcoming).

Smith, Robert W. *The Expanding Universe: Astronomy's "Great Debate," 1900-1931.* New York: Cambridge University Press, 1982.

Smith, Robert W. *The Space Telescope: A Study of NASA, Science, Technology, and Politics,* with contributions by Paul A. Hanle, Robert H. Kargon, and Joseph N. Tatarewicz. New York: Cambridge University Press, 1993, expanded and revised version.

Tucker, Wallace, and Karen Tucker. *The Cosmic Inquirers: Modern Telescopes and Their Makers.* Cambridge, Mass.: Harvard University Press, 1986.

van Helden, Albert. *Measuring the Universe: Cosmic Dimensions from Aristarchus to Halley.* Chicago: University of Chicago Press, 1985.

Wright, Helen. *Explorer of the Universe: A Biography of George Ellery Hale.* Woodbury, N.Y.: American Institute of Physics, 1994.

Wright, Helen, *The Great Palomar Telescope.* London: Faber and Faber, 1953.

PHOTOGRAPHY CREDITS

Cover, CORBIS; Original image courtesy of NASA/CORBIS; 1, NASA; 2-3, STS-82, Crew, HST,NASA; 4, Greg Bacon, Space Telescope Science Institute; 6, Don Foley; 8-9, NASA, The Hubble Heritage Team (STScI/AURA); 10, Bruce Balick (University of Washington), Vincent Icke (Leiden University, The Netherlands), Garrelt Mellema (Stockholm University), and NASA; 11, W.N. Colley and E. Turner (Princeton University), J.A. Tyson (Bell Labs, Lucent Technologies) and NASA; 12-14 (all), Courtesy David DeVorkin; 15, NASA, C.R. O'Dell and S.K. Wong (Rice University); 16-17, Anglo-Australian Observatory, Photography by David Malin; 18, NASA, Jayanne English (University of Manitoba), Sally Hunsberger (Pennsylvania State University), Zolt Levay (STScI), Sarah Gallagher (Pennsylvania State University), and Jane Charlton (Pennsylvania State University); 20-21 (both), Rare Books, Dibner Library/Smithsonian Institution; 23, Library of Congress; 25, J.P. Harrington and K.J. Borkowski (University of Maryland), and NASA; 26, Birr Scientific Heritage Foundation; 28-9, Robert Patterson, Stuart Levy, Donna Cox, National Center for Supercomputing Applications (NCSA)/Univ. of Illinois at Urbana-Champaign (UNIUC); David Malin, Anglo-Australian Observatory and Brent Tully, Univ. of Hawaii; 30, J. Bally, D. Devine, & R. Sutherland, D. Johnson (CITA), HST, NASA; 31, Carnegie Observatories, Carnegie Institution of Washington; 33, Margaret Bourke-White/gettyimages; 36, NASA, ESA, R. de Grijs (Institute of Astronomy, Cambridge, UK); 38-9, NASA, The Hubble Heritage Team (STScI/AURA); 40, NASA, Jeff Hester (Arizona State University); 41, NASA, Jeff Hester (Arizona State University); 42-3, E.J. Schreier (STScI) and NASA; 44-5 (both), NASA, The Hubble Heritage Team (STScI/AURA); 46-7, N.A. Sharp, REU Program/ NOAO/AURA/NSF; 48, California Institute of Technology; 51, Dana Berry (STScI); 52, Smithsonian Institution-National Air & Space Museum; 53, NASA, 55, Smithsonian Institution-National Air & Space Museum/Eric Long; 57, NASA; 58, Cal/Tech; 59, NASA; 60-1, NASA, ESA and Martino Romaniello (European Southern Observatory, Germany); 64, Andrea Dupree (Harvard-Smithsonian, CfA), Ronald Gilliland (STScI), NASA and ESA; 65, NASA; 66, S. Heap, NASA/Goddard Space Flight Center; 68-9, NASA; 70-1, NASA, The Hubble Heritage Team (STScI/AURA); 72, NASA, The Hubble Heritage Team (STScI/AURA); 73, NASA, ESA AND H.E. Bond (STScI); 74-5, NASA and G. Bacon (STScI); 76-77 (both), NASA, The Hubble Heritage Team (STScI/AURA); 78, NASA; 81, ESA, NASA and Robert A.E. Fosbury (ESA)/Space Telescope-European Coordinating Facility, Germany; 82, NASA; 85, NASA; 86, NASA, STScI; 89, NASA, Jeff Hester (Arizona State University); 90-1, NASA, 94, Ball Aerospace & Technologies Corp.; 95, NASA; 96-7, NASA, The Hubble Heritage Team (STScI); 98, NASA; 99, NASA, The Hubble Heritage Team (STScI/AURA); 100-1, Hubble Heritage Team (AURA/STScI/NASA); 102, NASA/J. Trauger (JPL); 103, NASA; 104-5, NASA; 106, T.A.Rector and Monica Ramirez/NOAO/AURA/NSF; 108, R. Gilmozzi, (STScI/ESA); Shawn Ewald, JPL; and

NASA; 110, NOAO/AURA/NSF; 111, NASA; 113, Science Museum/Science & Society Picture Library; 115-119 (all), NASA, The Hubble Heritage Team (STScI/AURA); 122, NASA, STScI; 123, John Bedke/STScI; 124, NASA, N. Walborn and J. Maiz-Apellániz (STScI, Baltimore, MD), R. Barbá (La Plata Observatory, La Plata, Argentina); 126-7, NASA, ESA, Jeff Hester (ASU); 128, NASA, The Hubble Heritage Team (STScI/AURA); 129, NASA/ESA; 130-1, NASA, NOAO, ESA, The Hubble Helix Nebula Team, M. Meixner (STScI) and T.A. Rector (NRAO); 132-133 (both), Jeff Hester and Paul Scowen (Arizona State University) and NASA; 134-5, NASA, ESA and The Hubble Heritage Team (STScI/AURA); 136, NASA, H. Ford (JHU), G. Illingworth (UCSC/LO), M. Clampin (STScI), G. Hartig (STScI), the ACS Science Team and ESA; 138, T.A. Rector & B.A. Wolpa, NOAO/AURA/NSF; 139, Jeff Hester and Paul Scowen (Arizona State University) and NASA; 141, NASA, The Hubble Heritage Team (STScI/AURA); 143,Geoffrey Clements/CORBIS; 145, NASA, The Hubble Heritage Team (STScI/AURA); 146, NASA, H. Ford (JHU), G. Illingworth (UCSC/LO), M.Clampin (STScI), G. Hartig (STScI), the ACS Science Team, and ESA; 148-149 (All), NASA, The Hubble Heritage Team (STScI/AURA); 152, NASA, ESA and D. Maoz (Tel-Aviv University and Columbia University); 155-157 (all), NASA, The Hubble Heritage Team (STScI/AURA); 158-9, NASA, J. Bell (Cornell Univ.) and M. Wolff (SSI); 160-1, C.R. O'Dell (Rice University) and NASA; 162, Raghvendra Sahai and John Trauger (JPL), the WFPC2 science team, and NASA; 163, Jon Morse (University of Colorado), and NASA; 164-166 (all), NASA, The Hubble Heritage Team (STScI/AURA); 169, Robert Williams and the Hubble Deep Field Team (STScI) and NASA; 170, NASA, The Hubble Heritage Team (STScI/AURA); 171, NASA, A. Fruchter and the ERO Team (STScI); 173, NASA, STScI; 174, NASA, H. Ford (JHU), G. Illingworth (UCSC/LO), M.Clampin (STScI), G. Hartig (STScI), the ACS Science Team, and ESA; 176, Henry E. Huntington Library and Art Gallery; 177, NASA, John Krist (STScI), Karl Stapelfeldt (JPL), Jeff Hester (Arizona State University), Chris Burrows (ESA/STScI); 178-9,Zolt Levay (STScI) and Randall Thompson (STScI); 180, Gary Bower, Richard Green (NOAO) and the STIS Instrument Definition Team, and NASA; 182, H. Hammel, (MIT), WFPC2, HST, NASA; 184, CORBIS, Original image courtesy NASA/CORBIS

Special thanks to Greg Bacon and the Space Telescope Science Institute for their generous contribution of the HST orbit diagram used on page 4 of this book.

ACKNOWLEDGMENTS

We have depended upon, and benefited from, many talented and dedicated people. Elizabeth Kessler wrote the entirety of Chapter 5 and lent her insight and expertise to the whole of the work. Marilyn Gibbons deftly succeeded in securing the best images, Paul Massoud provided thorough and critical fact checking, Nick Scoville, Ken Janes, John Huchra, and Karl Hufbauer kindly read portions of the manuscript. Zoltan Levay and the Hubble Heritage Team were most generous with their time and images. Trish Graboske, Joan Mathys, and Tracy McGowan provided advice and encouragement, Melissa Farris made the book visually appealing, and Johnna Rizzo kept us all on track. We also wish to acknowledge the support provided by the National Aeronautics and Space Administration, the National Science Foundation, and the National Air and Space Museum's Guggenheim fellowship program that led to historical research and exhibitry that informed the present work.

ABOUT THE AUTHORS

DAVID DEVORKIN is curator of the history of astronomy and the space sciences at the National Air and Space Museum, Smithsonian Institution, where he was curator for the Explore the Universe exhibition, opened in September 2001. His most recent books include *Beyond Earth: Mapping the Universe,* a companion volume to the exhibition; and *Henry Norris Russell: Dean of American Astronomers.*

ROBERT W. SMITH is Professor of History and past chair of the Department of History and Classics at the University of Alberta, and formerly a member of staff at the National Air and Space Museum. His books include the award-winning *The Space Telescope, A Study of NASA, Science, Technology and Politics,* and he has closely followed Hubble's history for over twenty years.

ELIZABETH KESSLER was a Guggenheim predoctoral fellow at the NASM during the writing of this work and is a graduate student in the History of Culture program at the University of Chicago where she is completing a dissertation on the aesthetics of Hubble images.

Index

THE HUBBLE SPACE TELESCOPE
Imaging the Universe

David DeVorkin and Robert W. Smith

PUBLISHED BY THE NATIONAL GEOGRAPHIC SOCIETY

John M. Fahey, Jr.   *President and Chief Executive Officer*
Gilbert M. Grosvenor   *Chairman of the Board*
Nina D. Hoffman   *Executive Vice President*

PREPARED BY THE BOOK DIVISION

Kevin Mulroy   *Vice President and Editor-in-Chief*
Charles Kogod   *Illustrations Director*
Marianne R. Koszorus   *Design Director*
Barbara Brownell Grogan   *Executive Editor*

STAFF FOR THIS BOOK

Johnna M. Rizzo   *Project Editor*
Melissa Farris   *Art Director*
Marilyn Mofford Gibbons   *Illustrations Editor*
Elizabeth Kessler   *Contributing Author*
John Paine   *Text Editor*
Sharon Kocsis Berry   *Illustrations Specialist*
Carl Mehler   *Director of Maps*
Paul Massoud   *Researcher*
Gary Colbert   *Production Director*
Lewis R. Bassford   *Production Project Manager*
Mark Wentling   *Indexer*

MANUFACTURING AND QUALITY CONTROL

Christopher A. Liedel   *Chief Financial Officer*
Phillip L. Schlosser   *Managing Director*
John T. Dunn   *Technical Director*
Maryclare McGinty   *Manager*

Composition for this book by the National Geographic Book Division.
Printed and bound by R. R. Donnelly & Sons, Willard, Ohio.
Color separations by Quad Imaging, Alexandria, Virginia.
Dust jacket printed by the Miken Co., Cheektowaga, New York.

One of the world's largest nonprofit scientific and educational organizations, the NATIONAL GEOGRAPHIC SOCIETY was founded in 1888 "for the increase and diffusion of geographic knowledge." Fulfilling this mission, the Society educates and inspires millions every day through its magazines, books, television programs, videos, maps and atlases, research grants, the National Geographic Bee, teacher workshops, and innovative classroom materials. The Society is supported through membership dues, charitable gifts, and income from the sale of its educational products. This support is vital to National Geographic's mission to increase global understanding and promote conservation of our planet through exploration, research, and education.

For more information, please call
1-800-NGS LINE (647-5463)
or write to the following address:

National Geographic Society
1145 17th Street N.W.
Washington, D.C. 20036-4688
U.S.A.

Visit the Society's Web site at www.nationalgeographic.com.

Library of Congress Cataloging-in-Publication Data

DeVorkin, David H., 1944-
  The Hubble Space Telescope / by David DeVorkin and Robert W. Smith.
    p. cm.
  Includes bibliographical references and index.
  ISBN 0-7922-6869-5 (regular) -- ISBN 0-7922-6892-X (deluxe)

  1. Hubble Space Telescope (Spacecraft)--History. 2. Space astronomy. I. Smith, Robert W. (Robert William), 1952- II. Title.
  QB500.268.D48 2004
  522'.2919--dc22

                        2004001150